Saving Our Planet

Saving Our Planet

Challenges and hopes

Mostafa K. Tolba

Executive Director
United Nations Environment Programme

CHAPMAN & HALL
London · New York · Tokyo · Melbourne · Madras

Published by Chapman & Hall, 2–6 Boundary Row, London SE1 8HN

Chapman & Hall, 2–6 Boundary Row, London SE1 8HN, UK

Chapman & Hall, 29 West 35th Street, New York NY10001, USA

Chapman & Hall Japan, Thomson Publishing Japan, Hirakawacho Nemoto Building, 6F, 1–7–11 Hirakawa-cho, Chiyoda-ku,
Tokyo 102, Japan

Chapman & Hall Australia, Thomas Nelson Australia,
102 Dodds Street, South Melbourne, Victoria 3205, Australia

Chapman & Hall India, R. Seshadri, 32 Second Main Road,
CIT East, Madras 600 035, India

First edition 1992

Typeset in 9.5/12 pt Bembo
Designed, produced and illustrated by Words and Publications, UK
Printed and bound in Hong Kong

ISBN 0 412 47370 4 (English), 0 412 47380 1 (French),
0 412 473909 (Spanish)

A catalogue record for this book is available from the British Library.

Library of Congress Cataloging-in-Publication data available

Contents

Preface

Man is both creature and moulder of his environment, which gives him physical sustenance and affords him the opportunity for intellectual, moral, social and spiritual growth. In the long and tortuous evolution of the human race on this planet a stage has been reached when, through the rapid acceleration of science and technology, man has acquired the power to transform his environment in countless ways and on an unprecedented scale. Both aspects of man's environment, the natural and the man-made, are essential to his well-being and to the enjoyment of basic human rights—even the right to life itself.

Stockholm Declaration, 1972

Mostafa K. Tolba

The environment is a complicated dynamic system, with many interacting components. Our knowledge of these components, of the interactions between them, and of the relationship between people, resources, environment and development has undergone profound evolution over the past two decades. We now realize that unless development is guided by environmental, social, cultural and ethical considerations, much of it will continue to have undesired effects, to provide reduced benefits or even fail altogether. Such 'unsustainable' development will only exacerbate the environmental problems that already exist. We all must come to terms with the reality of resource limitations and the carrying capacities of ecosystems. We must pursue plans that would not lead to conflicts over such limited resources and that would lead to what I called in 1974 'development without destruction'—or sustainable development that meets the needs of the present generation without compromising the ability of future generations to satisfy theirs.

This report analyses the changes (both positive and negative) that occurred in the environment since the convening of the United Nation Conference on the Human Environment in 1972 in Stockholm. It is based on published information available as of 30 November 1991. The report focuses not only on the state of the environment, but also on the interactions between development activities and environment. Both ultimately affect the human condition and human well-being.

The report is in five parts. The diagram overleaf illustrates the relationships between these parts, and should help the reader

... preface

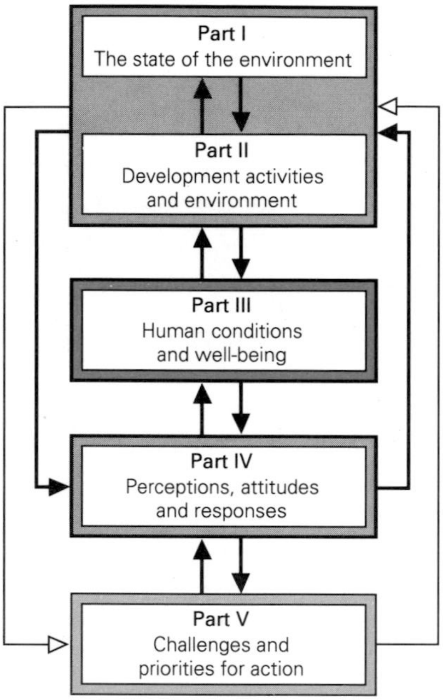

Part I
The state of the environment

Part II
Development activities and environment

Part III
Human conditions and well-being

Part IV
Perceptions, attitudes and responses

Part V
Challenges and priorities for action

understand the complex interactions involved.

The question now is 'where do we go from here?' First, I must emphasize that the goal of sustainable development cannot be attained without significant changes in the ways development initiatives have been planned and implemented. These changes will not come about unless there are similar changes in everybody's perceptions of environmental issues and attitudes towards them; the public, governments, business and industry. They will not come about if we do not stop taking the environment and its natural resources for granted, stop considering it all as free goods. They will not come about if we do not consider environmental protection and environmental security as essential parts of national and international security. Secondly, environmentally sound development plans will not succeed without public participation and a sense of individual responsibility. And finally, it has become more evident than ever in the past two decades that environmental problems are not restricted to national boundaries but that most of them are of regional and global significance. Thus international cooperation—global partnership—is essential not only to protect the environment, but also to set the world on the path to sustainable development. The challenges ahead are formidable. To translate good intentions and high-sounding declarations into action is long overdue.

Mostafa K. Tolba
Executive Director
United Nations Environment Programme

Nairobi, January 1992

Acknowledgements

I am extremely grateful to my colleague Professor Essam El-Hinnawi of the Egyptian National Research Centre for his outstanding efforts in putting all this material together, and for his unfailing efforts in checking and rechecking the data and facts presented in this publication.

Thanks are also due to Robin Clarke, of Words and Publications, who spared no effort in polishing up the sometimes dry scientific language of the text, and to Nigel Jones, also of Words and Publications, for designing the publication in a readable and attractive form.

Photo credits

Page x	UNEP
Page 12	Image Bank
Page 32	Dylan Garcia, Panos Pictures
Page 44	FAO
Page 56	FAO
Page 62	UNEP
Page 70	M.D. Gwynne
Page 80	Jeremy Hartley, Panos Pictures
Page 104	UNEP
Page 118	FAO
Page 136	F. Botts, FAO
Page 160	Doug Hulcher, Panos Pictures
Page 168	Doug Hulcher, Panos Pictures
Page 176	F. Botts, FAO
Page 186	M. Jones, FAO
Page 196	A. Girod, FAO
Page 208	J. Young, Panos Pictures
Page 220	Jeremy Hartley, Panos Pictures
Page 230	Wind Energy Group Ltd

Part I

The State of the Environment

Chapter 1

Atmospheric pollution

Atmospheric pollution is a major problem facing all nations. It is caused by chemicals emitted into the air from both natural and man-made sources. Emissions from natural sources include those from living and non-living sources (such as plants, radiological decomposition, forest fires, volcanic eruptions, and emissions from land and water). These emissions lead to natural background pollution levels that vary with the local source of emission and the prevailing weather. People have caused air pollution since they learned how to use fire, but man-made (anthropogenic) air pollution has increased rapidly since industrialization began.

Research over the past two decades has revealed that in addition to the common air pollutants—sulphur oxides, nitrogen oxides, particulate matter, hydrocarbons and carbon monoxide—many volatile organic compounds and trace metals are emitted into the atmosphere by human activities. Although our knowledge of the nature, quantity, behaviour and effects of air pollutants has greatly increased in recent years, more needs to be known about the fate and transformation of different pollutants and about their combined (synergistic) effects on human health and the environment.

Worldwide, 99 million tonnes of sulphur oxides (SO_x), 68 million tonnes of nitrogen oxides (NO_x), 57 million tonnes of suspended particulate matter (SPM), and 177 million tonnes of carbon monoxide (CO) were released into the atmosphere in 1990 as a result of human activities, from stationary and mobile sources (1). The countries of the Organization for Economic Cooperation and Development (OECD) accounted for about 40 per cent of the SO_x, 52 per cent of the NO_x, 71 per cent of the CO, and 23 per cent of the SPM emitted into the global atmosphere (Figures 1.1 and 1.2).

Figure 1.1
Man-made emissions of common air pollutants

based on data from (1, 32, 33)

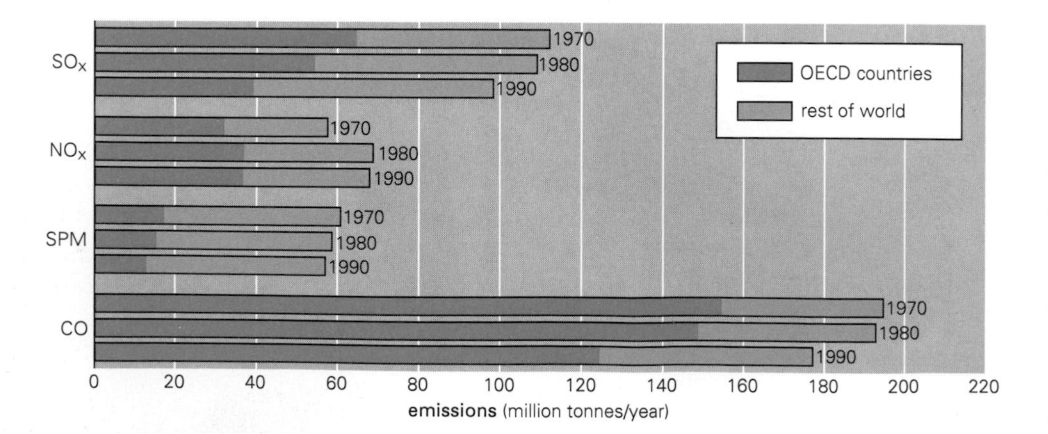

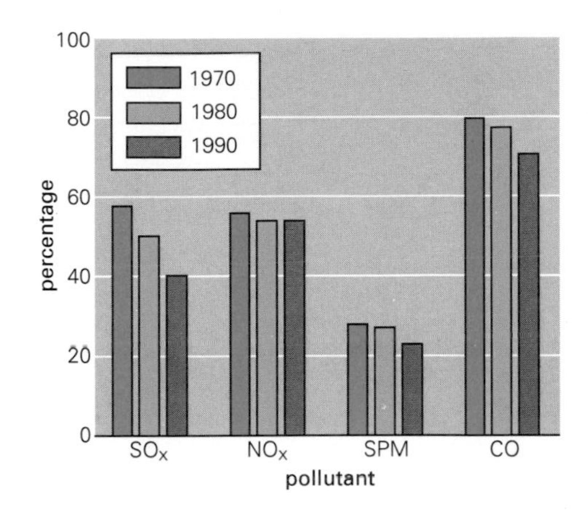

Figure 1.2
Contribution of OECD countries to global air emissions

based on data from (1, 32, 33)

Time-series data (Figure 1.1) show that although levels of SO_x emissions peaked in 1970 at a high of about 115 million tonnes, they dropped to 99 million tonnes in 1990 as a result of a marked reduction in SO_x emissions in OECD countries (Figure 1.2). These reductions have been achieved mainly by stricter regulations of emissions, changes in energy structures and fuel prices, and the introduction of more efficient technologies. Between 1970 and 1990, SO_x emissions in the OECD countries decreased from about 65 to 40 million tonnes. In contrast, SO_x emissions in the rest of the world increased from 48 to 59 million tonnes over the same period. From 1970 to 1990, there were no marked changes in NO_x and SPM emissions. There was, however, a marked decrease in CO emissions in the OECD countries, from 155 to 125 million tonnes; in the rest of the world CO emissions increased from about 40 to 52 million tonnes between 1970 and 1990, mainly due to the increase in automobile traffic.

In the past two decades, and especially in the 1980s, increasing attention has been given to the emission into the atmosphere of hundreds of trace compounds—organic and inorganic. Some 261 volatile organic chemicals (VOCs) have been detected in ambient air (2). In most cases, the concentrations are low—less than one part per billion by volume (ppbv). Some of these VOCs are highly reactive, even at such low concentrations, and are suspected of playing an important role in the formation of photochemical oxidants. Another group of compounds that has recently received attention is trace metals such as cadmium, copper, mercury and zinc (Figure 1.3). Lead is the best studied of these metals. An estimated 80-90 per cent of lead in ambient air derives from the combustion of leaded petrol (Chapter 14).

Because of growing concern about air pollution, programmes were initiated in some developed countries in the 1960s to monitor the common pollutants and assess changes in air quality. In 1973, the World Health Organization (WHO) set up a global programme to help countries

Between 1970 and 1990, SO_x emissions in the OECD countries decreased from about 65 to 40 million tonnes. In contrast, SO_x emissions in the rest of the world increased from 48 to 59 million tonnes over the same period.

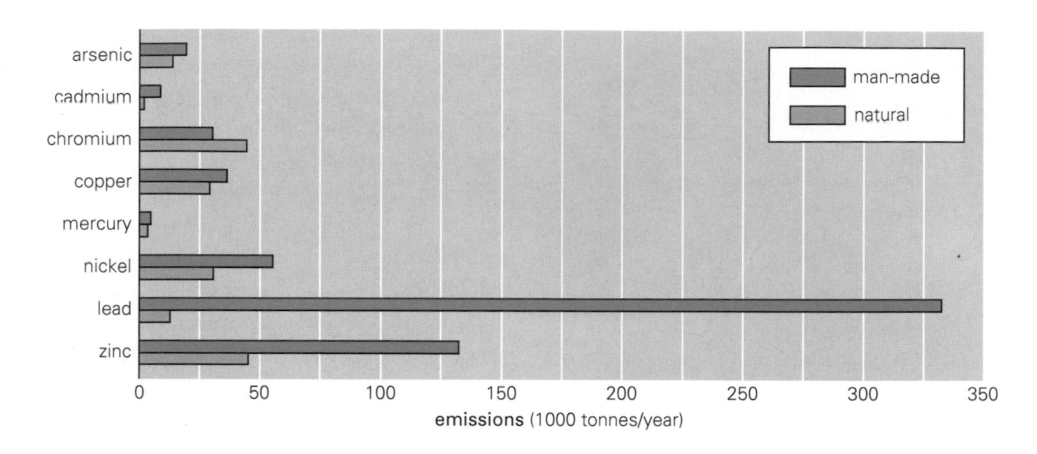

Figure 1.3 Global natural and man-made emissions of selected trace elements (1983)

based on data from (34, 35)

monitor air pollution. This project became a part of UNEP's Global Environment Monitoring System (GEMS) in 1976. Some 50 countries now participate in the GEMS/AIR monitoring project, and data are obtained at about 175 sites in 75 cities, 25 of them in developing countries.

Data from GEMS/AIR during 1980-84 indicate that of 54 cities 27 (including Auckland, Bucharest, Bangkok, Toronto and Munich) have acceptable air quality with SO_2 concentrations below 40 micrograms/cubic metre ($\mu g/m^3$); WHO has established a range of 40–60 $\mu g/m^3$ as a guideline for exposure to avoid increased risk of respiratory diseases. Eleven cities (including New York, Hong Kong and London) have marginal air quality, with SO_2 concentrations of 40–60 $\mu g/m^3$. The other 16 cities (which include Rio de Janeiro, Paris and Madrid) have unacceptable air quality, with SO_2 concentrations exceeding 60 $\mu g/m^3$ (3, 4). Data for 41 cities indicate that 8 (including Frankfurt, Copenhagen and Tokyo) have acceptable SPM concentrations below 60 $\mu g/m^3$ (the WHO range is 60–90 $\mu g/m^3$). Ten cities (including Toronto, Houston and Sydney) have borderline SPM concentrations of 60–90 $\mu g/m^3$, and 23 (including Rio de Janeiro, Bangkok and Tehran) have SPM concentrations exceeding 90 $\mu g/m^3$. The extraordinary levels noted in some cities in developing countries can be partially explained by natural dust; other culprits include the black, particulate-laden smoke spewed out by diesel-fueled vehicles lacking even rudimentary pollution control. The GEMS/AIR assessment concluded that nearly 900 million people living in urban areas around the world are exposed to unhealthy levels of SO_2 and more than one billion people are exposed to excessive levels of particulates.

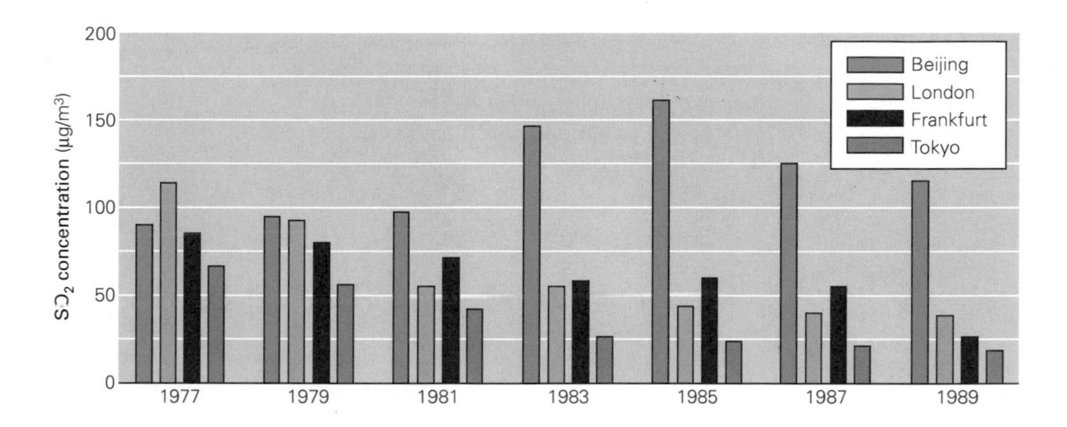

Figure 1.4
Sulphur dioxide levels in the air of selected cities (µg/m³)

based on data from (16)

In some cities, such as Tokyo, Frankfurt and London, air quality has improved, with a marked fall in the average annual concentration of SO_2 and in the number of days in which air quality guidelines are exceeded. However, in several cities—especially in the developing countries—the SO_2 concentration still exceeds the WHO guideline (Figure 1.4). Levels of suspended particulate matter also declined in most OECD cities in the 1970s, and have since levelled off (1). On the other hand, there has been no marked trend in the concentration of ambient NO_x over the past two decades.

Ozone and other photochemical oxidants such as peroxyacetyl nitrate (PAN) are formed in the lower atmosphere from NO_x and hydrocarbon emissions in the presence of sunlight during stagnant high-pressure weather conditions. These occur most often in the summer, and lead to the photochemical smog that is characterized by a thick layer of brown haze. Ozone concentrations in OECD countries, where time-series data are available, have not shown a clear trend; the principal reason is that they depend largely on prevailing weather conditions, which can change considerably from year to year. In many OECD countries, ozone levels exceed recommended standards. In the United States, the limit of exposure of 235 µg/m³ (for a maximum of one hour a day) is often exceeded, and it has been estimated that some 75 million people are exposed to higher levels of ozone (1, 5).

The GEMS/AIR assessment concluded that nearly 900 million people living in urban areas around the world are exposed to unhealthy levels of SO_2 and more than one billion people are exposed to excessive levels of particulates.

Ozone has long been considered to be the oxidant that determines the air quality of an urban atmosphere. During the 1980s, however, atmospheric chemists identified hydrogen peroxide, a photochemical product in the air, as another oxidant that may significantly degrade air quality (6). Measurements of hydrogen peroxide carried out at various locations in Brazil, Canada, Europe, Japan and the United States show concentrations generally less than 10 ppbv. No guidelines have yet been established for exposure to ambient hydrogen peroxide.

Air pollution is not only restricted to the outdoor environment. Although indoor air pollution has been known since prehistoric times, it has recently become a matter of concern. High concentrations of indoor air pollutants continue to be a fact of life for people who cook over open fires fuelled by charcoal, coal, wood, dung and agricultural residues. The expression 'sick building syndrome' has been used to describe buildings in which the air causes symptoms such as eye, nose and throat irritation, mental fatigue, headache, nausea, dizziness, airway infections, and dry mucous membranes. Such symptoms have been epidemiologically related to sealed buildings, non-openable windows, tight-enclosure dwellings, increased temperature and dust levels, and passive cigarette smoking (7, 8).

Indoor air pollution in residences, public buildings and offices is created mainly by the occupants' activities and their use of appliances, power equipment and chemicals; by emissions from structural and decorative material; by thermal factors; and by the penetration of outdoor pollutants (9, 10, 11, 12). The most important indoor contaminants are tobacco smoke, radon decay products, formaldehyde, asbestos fibres, combustion products (such as NO_x, SO_x, CO, CO_2 and polycyclic aromatic hydrocarbons), and other chemicals arising from household activities. WHO (7) indicated that several microbiological air contaminants are encountered in the indoor environment. These include moulds and fungi, viruses, bacteria, algae, pollens, spores and their derivatives. Recently, more than 66 volatile organic chemicals have been identified in indoor air (2, 13). Several studies (11, 12, 14) have pointed out that many pollutants are more concentrated in the indoor environment than they are outside. Levels of respirable particulate matter, NO_x, CO_2, CO, formaldehyde, radon and several other materials are higher indoors than outdoors.

Pollutants emitted into the atmosphere do not necessarily remain confined in the area near the source of emission or the local environment. They can be transported over long distances, cross frontiers, and create regional and global environmental problems.

Acidic deposition is one of these problems (for ozone depletion and the effect of greenhouse gases, see Chapters 2 and 3). Worldwide monitoring of precipitation has established that precipitation in extensive areas of North America and Europe is about 10 times more acidic than normal (15). Large-scale acidification due to man-made emissions of sulphur and nitrogen is not considered to be a significant problem in other regions at present (16). However, there are indications that certain tropical regions, for example, south-eastern Brazil, southern China, south-western India, Jamaica, northern Venezuela and Zambia may experience problems relating to acidification if current trends in urbanization and industrialization continue into the 21st century (15).

The mechanisms by which the emitted pollutants, mainly SO_2 and NO_x, are transformed into acidifying substances in both the gaseous and liquid phases are complex and incompletely understood. The concentration and distribution of acidic deposits, wet and dry, are determined by many interacting processes such as the transport and dispersal of the parent pollutants, the role of oxidizing agents such as hydrocarbon derivatives and ozone, and meteorological factors. The basic physical and chemical processes involved in the washout of soluble gases and aerosol species are not limited to SO_x and NO_x. Most atmospheric trace gases are highly soluble in precipitation. In fact, measurements have recently shown that precipitation contains hundreds of organic compounds (I7) and many trace metals (18, 19). Acid fog recently reported in the United States and other countries contains carbonyl compounds, alkyl sulphonate and pesticides in addition to sulphates and nitrates (20, 21, 22, 23).

Measurement of precipitation chemistry on a global scale is conducted as part of the work of the World Meteorological Organizition's (WMO) Background Air Pollution Monitoring Network (BAPMoN). First established in 1969, the BAPMoN network, currently a WMO/UNEP joint effort, comprises some 196 stations, 152 of which are equipped to measure precipitation chemistry. Data from BAPMoN show that during 1972–84 global levels of sulphur in precipitation fell (4). This trend is due to the decrease in global emissions of SO_x (Figure 1.1).

On a regional basis, the Cooperative Programme for

Worldwide monitoring of precipitation has established that precipitation in extensive areas of North America and Europe is about 10 times more acidic than normal.

Monitoring and Evaluation of Long-range Transmission of Air Pollution in Europe (EMEP) was established in 1977 as a joint UN-ECE/WMO/UNEP venture. EMEP is responsible for coordination of routine measurements of air and precipitation quality at a network of 102 sites located throughout Europe. A number of EMEP stations are also BAPMoN stations. Recent results from EMEP show that much of central and eastern Europe receives rainfall containing sulphate in excess of 1 milligram of sulphur per litre of precipitation (mg S/l). The highest concentrations, more than 1.5-2.0 mg S/l, are recorded in eastern Europe. However, during 1983–87 the size of the area receiving precipitation with sulphate concentrations of more than 1.5 mg S/l was reduced compared with that for the period 1978–82 (16), mainly because of reduction in SO_x emissions. Nitrate levels in precipitation are greatest over the Baltic Sea, eastern Germany and northern Poland. The concentration of ammonia in precipitation is high over parts of Belgium, France and The Netherlands, and also over an area in eastern Europe near the Poland/Czechoslovakia/Soviet border.

Impacts of atmospheric pollution

Air pollution affects human health, vegetation and various materials. The notorious sulphurous smog that occurred in London in 1952 and 1962, and in New York in 1953, 1963 and 1966, demonstrated the link between excessive air pollution and mortality and morbidity (Chapter 18). Such acute air pollution episodes occur from time to time in some urban areas. In January 1985, one such episode occurred throughout western Europe. Near Amsterdam, the 24-hour average SPM and SO_x concentrations were each in the range of 200-250 $\mu g/m^3$—much higher than the WHO guideline values. During the episode several thousand people were affected; pulmonary functions in children were 3 to 5 per cent lower than normal, and this dysfunction persisted for about 16 days after the episode (24). Air pollution episodes of this type occur frequently in Athens. But even in the absence of such episodes, long term exposure to air pollution can affect several susceptible groups—the elderly, children and those with respiratory and heart conditions.

Air pollution can cause substantial damage to many materials (25); many historical buildings and monuments have now been severely damaged by air pollution, mainly SO_x. The Acropolis in Greece, the Colosseum in Italy and the Taj Mahal in India withstood the influence of the atmosphere for thousands of years without any great damage. Yet in the past few decades their surfaces have been severely damaged by air pollution.

Indoor air pollution has a number of effects. Reference has already been made to the sick building syndrome, which causes a substantial portion of disease and absenteeism from work or school (8). Recently, attention focussed on the possible health hazards of radon emissions at home. In the United States, it has been found that the concentration of radon indoors is about six times higher than that outdoors (26), and that the current annual mortality rate from lung cancer attributable to indoor radon exposure is about 16 000 cases. However, it was found that only three per cent of this mortality occurred among individuals who never smoked tobacco. Thus more than 90 per cent of the lung cancer risk associated with radon could be controlled by eliminating smoking. The penetration of outdoor pollutants into buildings has also been a cause of concern. High ozone levels have been found in some museums and art galleries, and there are fears that ozone—a highly reactive gas—could cause the fading of colours of works of art. Several museums and art galleries have taken costly precautions to monitor ozone levels indoors, and to isolate paintings and other works of art.

Emissions from burning biomass fuels, especially in rural areas of developing countries, are a major source of indoor air pollution. The most important identified adverse effects are chronic obstructive pulmonary disease and nasopharyngeal cancer (27, 28). When infants are exposed to such pollution, acute bronchitis and pneumonia occur because respiratory defences are impaired. Emissions from biomass and coal burning at home contribute significantly to outdoor air pollution in some areas. Indoor emissions create a haze in certain parts of the Himalayas which may have effects on visibility and on vegetation in that mountain ecosystem (14).

Considerable evidence has accumulated over the past two decades to show that acidic deposition poses a threat to various resources: lakes and their aquatic life, forestry, agriculture and wildlife (29). Thousands of lakes in parts of Scandinavia, the north-east United States, south-east Canada and south-west Scotland have been affected by acidic deposition to varying degrees, and many of the lakes (especially in Sweden and Norway) have lost their fish partly

Thousands of lakes in parts of Scandinavia, the north-east United States, south-east Canada and south-west Scotland have been affected by acidic deposition ... and many (especially in Sweden and Norway) have lost their fish partly or totally.

or totally. Acidic deposition has also caused excessive leaching of some trace metals from soils and the bottom sediments of lakes, resulting in high concentrations of these elements in lakes and groundwater. The effects of acidic deposition on the degradation of forests in Europe and in North America are also well documented (Chapter 7).

Responses

Although it was thought that urban (and rural) air pollution problems are local problems, it has become increasingly evident that urban emissions lead to the regional and global distribution and deposition of pollutants. These scales are not isolated from one another, and solutions to problems on one scale may lead to new problems in another. For example, the use of tall stacks to disperse pollutants may abate local air pollution, but it can lead to the dispersion—and eventual deposition—of primary pollutants and their reaction products on a regional and even global scale. It therefore became evident in the past two decades that countries have to work in concert to reduce air pollution.

Several countries have had marked success in reducing emissions into the atmosphere by implementing stricter control regulations, switching to low-sulphur fuels, and installing air pollution control equipment at enterprises. In Bulgaria, for example, emissions of suspended particulate matter were reduced by 1.6 million tonnes a year during the period 1976–80 (29). Comparable reductions in air pollutants have been recorded in OECD countries (1) and in a few developing countries (such as Singapore). An indicator of the efforts to reduce air pollution is the growth in sales of air pollution control equipment such as flue gas desulphurization equipment and electrostatic precipitators. A recent market survey (30) indicates that total orders for such equipment worldwide reached $12.7 billion in 1991 ($4.0 billion in North America, $4.2 billion in Europe, and $4.5 billion for the rest of the world). This is more than double the figure for a decade ago.

The signing of the UN-ECE Convention on Long-range Transboundary Air Pollution in 1979 demonstrated the determination of European and North American countries to work together to cut back sulphur and nitrogen oxide emissions to acceptable levels. In 1987, the Protocol to the Convention on the Reduction of Sulphur Emissions or their Transboundary Fluxes—by at least 30 per cent from 1980 levels by 1993—entered into force. The Protocol Concerning the Control of Emissions of Nitrogen Oxides or their Transboundary

Fluxes, signed in November 1988, calls for a freeze on emissions at 1987 levels in 1994, as well as discussions set to begin in 1996 aimed at actual reductions. Some countries have made commitments to go beyond both protocols. At least nine have pledged to bring SO_2 levels down to less than half of 1980 levels by 1995. Austria, Sweden and Germany have committed themselves to reducing SO_2 levels by two-thirds. For NO_x, 12 Western European nations have agreed to go beyond the freeze and reduce emissions by 30 per cent by 1998. A November 1988 directive by the European Economic Community represents a binding commitment by its members to reduce significantly the emissions that cause acid rain. The directive will lower community-wide emissions of SO_2 from existing power plants by a total of 57 per cent from 1980 levels by 2003, and of NO_x by 30 per cent by 1998 (5).

To rehabilitate areas acidified by acidic deposition, liming programmes have been undertaken in some countries, especially in Sweden. More than 3000 lakes have been limed in Sweden since 1976 (31). Liming is also used to reduce the acidity of soils in forests. However,the most effective way of preventing acidification is to reduce emissions at source. Several technologies are now available to reduce the sulphur content of coal and to eliminate SO_x emissions from stack gases (Chapter 13).

The regulation of indoor air quality is much more complex than the regulation of outdoor air quality. Outdoor air is a public good in the sense that members of a community breathe the same ambient air. The rationale for government regulation of outdoor air pollution is the protection of the health of the members of the community on an equal basis. This situation is quite different for some indoor environments, especially private residences. If occupants foul the air in their own home, they are forced to breath it. If they attempt to improve its quality, by increasing ventilation for example, they bear the costs and enjoy the benefits. The problem of regulating indoor air quality is, therefore, highly dependent on public perception and awareness of the different risks involved.

An indicator of the efforts to reduce air pollution is the growth in sales of air pollution control equipment ... A recent market survey indicates that the total orders for such equipment worldwide reached $12.7 billion in 1991.

Chapter 2

Ozone depletion

Figure 2.1
Estimated world consumption (top) and sectoral use (bottom) of the main CFCs and halons (1986)

based on data from (19)

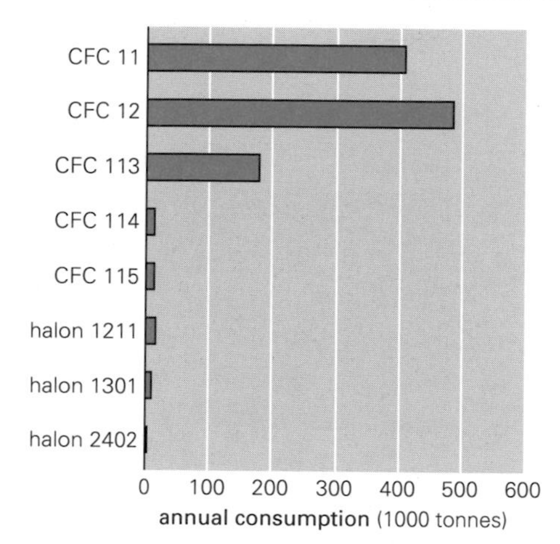

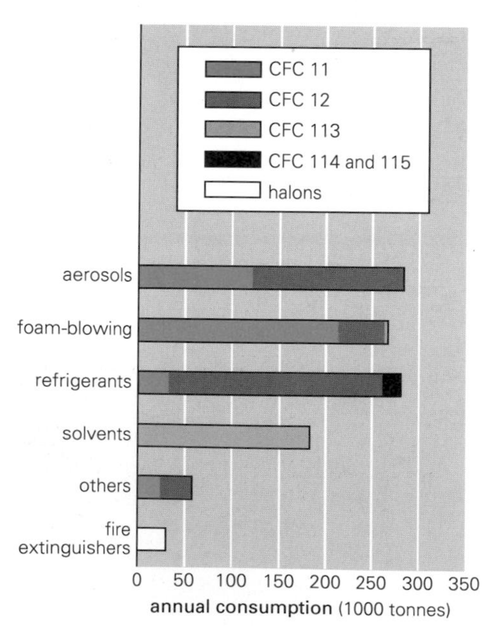

In contrast to the harmful ozone formed as a photochemical oxidant at ground level (tropospheric ozone, see Chapter 1), ozone in the stratosphere, between 25 and 40 km above the earth's surface, is the natural filter that absorbs and blocks the sun's short wavelength ultraviolet radiation (UV-B) that is harmful to life.

Ozone exists in equilibrium in the stratosphere, balanced between formation from molecular oxygen and destruction by ultraviolet radiation. The presence of reactive chemicals in the stratosphere, such as the oxides of hydrogen, nitrogen and chlorine, can accelerate the process of ozone destruction and therefore upset the natural balance, leading to a net reduction of the amount of ozone. These chemicals can participate in many ozone-destroying reactions before they are removed from the stratosphere.

Concern about the depletion of stratospheric ozone by human activities dates from the late 1960s over emissions of nitrogen oxides (NO_x) by high-flying supersonic aircraft. The high temperatures of the engines convert atmospheric nitrogen and oxygen into NO_x and deposit them in the stratosphere at flight altitudes of 17 to 20 km. The NO_x then act as a catalyst destroying ozone in the stratosphere (1, 2). Later, in 1974, it was found that man-made chlorofluorocarbons (CFCs), although inert in the lower atmosphere, can survive for many years and migrate into the stratosphere. There, CFCs are destroyed by ultraviolet radiation, releasing atomic chlorine which attacks stratospheric ozone, forming the free radical ClO^- which reacts further to regenerate atomic chlorine. This chain reaction can cause the destruction of as many as 100 000 molecules of ozone per single atom of chlorine (3, 4, 5, 6, 7).

Chlorofluorocarbons are compounds used as propellants and solvents in aerosol sprays; fluids in refrigeration and air-conditioning equipment; foam blowing agents in plastic foam production; and solvents, mainly in the electronics industry. Although there is a range of compounds called chlorofluorocarbons, CFC 11 (trichlorofluoromethane) and CFC 12 (dichlorodifluoromethane) are the most commonly used (Figure 2. 1). Studies in the 1980s have shown that emissions of bromine can also lead to a significant reduction in stratospheric ozone (8). Bromofluorocarbons (halons 1211 and 1301) are widely used in fire extinguishers, and ethylene dibromide and methyl bromide are used as fumigants.

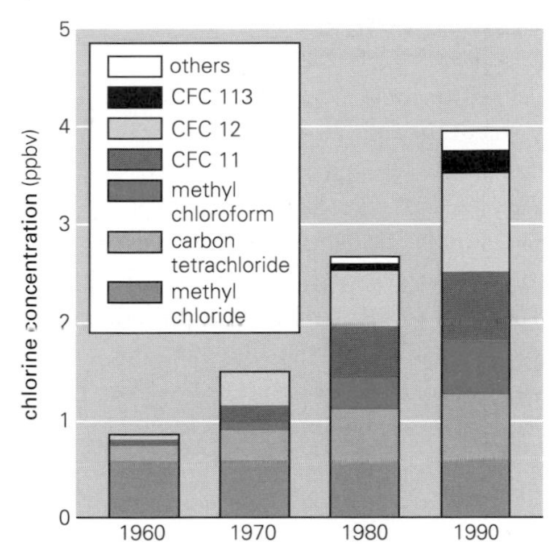

Figure 2.2 (above) Organochlorine concentration in the atmosphere

based on data from (7)

The concentration of chlorine in the stratosphere is determined mainly by anthropogenic sources of CFC 11, CFC 12, carbon tetrachloride and methyl chloroform. Methyl chloride is the only natural organochlorine compound found in the atmosphere. The concentration of chlorine in the atmosphere due to methyl chloride has remained unchanged perhaps since 1900. The major additions of chlorine to the atmosphere occurred mainly since 1970 and have been attributed to anthropogenic sources (Figure 2.2). At present the total chlorine in the atmosphere due to organochlorine compounds is approaching 4.0 ppbv—an increase by a factor of 2.6 in only 20 years.

A chemical's ability to destroy ozone depends on a combination of the percentage of chlorine released from it by ultraviolet radiation and its lifetime in the atmosphere. This determines its ozone depletion potential (ODP). ODP is measured relative to CFC 11, which has been given an ODP of 1.0 (Figure 2.3).

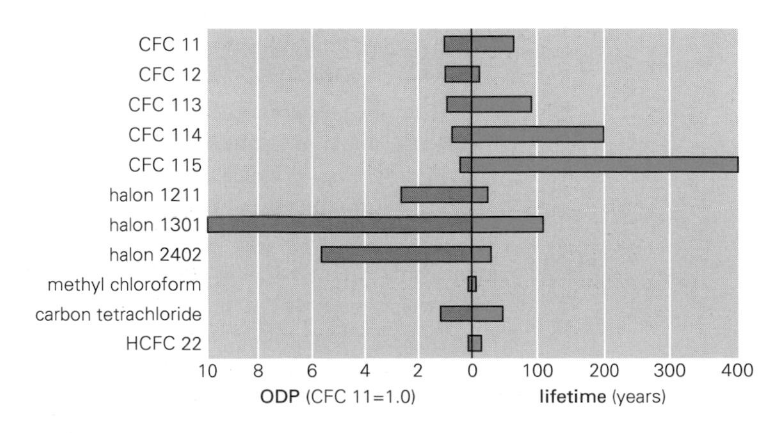

Figure 2.3 (right) Ozone depletion potential and lifetime of halocarbons

based on data from (19, 24)

Has the ozone layer been depleted?

Observing changes in the chemistry of the stratosphere caused by the release of trace gases is difficult. Theoretical models have therefore been developed to estimate these changes. The models developed in the 1970s estimated that the release of chlorofluorocarbons at late 1970s rates, if continued indefinitely, would deplete stratospheric ozone by about 15 per cent, with an uncertainty range of 6 to 22 per cent (9, 10, 11). Later models (12, 13, 14) indicate that if production of CFCs were to continue into the future at the 1980 rate, the steady state reduction in total global ozone could be about or less than 3 per cent over the next 70 years. If the release rate of CFCs should become twice the level of 1980, or if stratospheric chlorine reaches 15 ppbv, it has been predicted that there will be a 3–12 per cent reduction of the ozone column, assuming that the annual rates of increase in the atmospheric concentration of carbon dioxide, nitrous oxide and methane continue at their present rate. There are several limitations to these theoretical models, and they may be underestimating the adverse impact of CFCs on ozone, especially at high latitudes in winter (15).

Worldwide ozone monitoring (ground-based total ozone monitoring) began during the International Geophysical Year in 1957, but only a few stations have continuous records from 1957 to the present day. One of these is in Halley Bay, Antarctica. Records from Halley Bay showed that the total ozone levels above the station in 1984 were only about 60 per cent of those obtained in the late 1950s and early 1960s (16). The changes were most pronounced in October. Recent studies (15, 17, 18) have indicated an average decrease of 30–40 per cent in the total column of ozone in the lower stratosphere between 15 and 20 km above Antarctica (referred to as the ozone hole); at some altitudes the ozone loss may be as high as 95 per cent (18). The data indicate that the decrease in ozone occurs in springtime (September–October) and recovers in summer (January–February).

Although different theories have been put forward to explain the massive Antarctic ozone depletion (6), scientific evidence strongly indicates that man-made chlorinated and brominated chemicals are primarily responsible for this depletion. Under the special meteorological conditions of the Antarctic winter stratosphere (very low temperatures and abundant polar stratospheric clouds) chlorine and nitrogen interact to produce massive ozone depletion in the lower stratosphere when sunlight returns in the spring. Similar conditions have also been identified in the Arctic stratosphere. However, ozone damage over the Arctic is not yet comparable to that over the

Antarctic. The degree of any future ozone depletion will probably depend on the particular meteorology of each Arctic winter, and future atmospheric levels of chlorine and bromine.

An analysis of the total column ozone data from ground-based instruments has shown measurable downward trends from 1969 to 1988 of 3 to 5.5 per cent in the northern hemisphere, between latitudes 30 and 64 °N (19). Recent analyses of TOMS (Total Ozone Mapping Spectrometer) satellite data by NASA indicate that total ozone between 65 °N and 65 °S has been decreasing at an average rate of about 0.26 per cent per year. Statistically meaningful ozone depletions of 3 to 5 per cent are indicated north of 35 °N in springtime and may reach 9 per cent at 45 °N in winter. A 4 per cent or greater ozone depletion is indicated at all latitudes south of about 40 °S throughout the year. No statistically significant ozone trends were found between 30 °N and 30 °S throughout the year (20). While the cause of the observed ozone depletion has not been unequivocally identified, the ozone changes coupled with other atmospheric data are strongly suggestive of a chlorine-induced effect.

Impacts of ozone depletion

The depletion of the ozone layer will increase the intensity of ultraviolet (UV-B) radiation reaching the earth's surface. It has been predicted that a one per cent reduction in the amount of stratospheric ozone will lead to an increase of approximately 2 per cent in UV-B radiation reaching the ground. UV-B radiation is known to have a multitude of effects on humans, other animals, plants and materials. Most of these effects are damaging.

Exposure to increased UV-B radiation can cause suppression of the body's immune system, which might lead to an increase in the occurrence or severity of infectious diseases such as herpes, leishmaniasis and malaria, and a possible decrease in the effectiveness of vaccination programmes. Enhanced levels of UV-B radiation can lead to increased damage to the eyes, especially cataracts. Each 1 per cent decrease in total column ozone is expected to lead to an increase of 0.6 per cent in the incidence of cataracts (or an estimated worldwide increase of 100 000 blind persons per year due to

Each 1 per cent decrease in total column ozone is expected to lead to an increase of 0.6 per cent in the incidence of cataracts (or an estimated world-wide increase of 100 000 blind persons per year due to UV-B induced cataracts).

UV-B induced cataracts). In addition, every 1 per cent decrease of total column ozone is predicted to lead to a 3 per cent rise of the incidence of non-melanoma skin cancer (or an estimated worldwide increase of 50 000 cases per year). There is also concern that an increase of the more dangerous cutaneous malignant melanoma could also occur. A recent study has shown that a 1 per cent reduction in ozone will result in a 1.6 per cent increase in male death rates and a 1.1 per cent increase in female death rates due to melanoma (21).

Plants vary in their sensitivity to UV-B radiation. Some crops, such as peanut and wheat, are fairly resistant while others, such as lettuce, tomato, soybean and cotton, are sensitive. UV-B radiation alters the reproductive capacity of some plants and also the quality of harvestable products. This could have serious effects on food production in areas of already acute shortage (19).

Increased UV-B radiation has negative effects on aquatic organisms, especially small ones such as phytoplankton, zooplankton, larval crabs and shrimp, and juvenile fish. Because many of these small organisms are at the base of the marine food web, increased UV-B exposure may have negative effect on the productivity of fisheries. Increased levels of UV-B radiation may also modify freshwater ecosystems by destroying micro-organisms,thus reducing the efficiency of natural water purification.

Substantial reductions in upper stratospheric ozone and associated increases in ozone in the lower stratosphere and upper troposphere might lead to undesirable global perturbations in the earth's climate. The vertical redistribution of ozone may warm the

The Montreal Protocol

The Montreal Protocol on Substances that Deplete the Ozone Layer was adopted in September 1987. The Protocol entered into force on 1 January 1989. As of 31 August 1991, 73 countries and the EEC had become parties to the protocol.

The controls

CFC 11, CFC 12, CFC 113, CFC 114 and CFC 115

- As of 1 July 1989, and within 12 months, and thereafter, the level of consumption and production should not exceed the 1986 level.
- As of 1 July 1993, and within 12 months, and thereafter, the level of consumption and production should not exceed 80 per cent of the 1986 level.
- As of 1 July 1998, and within 12 months, and thereafter, the level of consumption and production should not exceed 50 per cent of the 1986 level.

Halons 1301, 1211 and 2402

- As of 1 February 1992, and within 12 months, and thereafter, the level of consumption and production should not exceed the 1986 level.

lower atmosphere and reinforce the greenhouse effect associated with an increase in carbon dioxide. In addition, chlorofluorocarbons are among the potential greenhouse gases (see Chapter 3).

Responses

Ozone depletion is a global problem that requires global action. International efforts, coordinated and catalysed by UNEP since 1977, in full co-operation with WMO, the scientific community and industry, led to the development of the Vienna Convention for the Protection of the Ozone Layer which was adopted in March 1985. The purpose of the convention was to promote information exchange, research and systematic observations to protect human health and the environment against adverse effects resulting or likely to result from human activities that modify or are likely to modify the ozone layer. The convention was designed so that protocols requiring specific control measures could be added, and in September 1987 the Montreal Protocol on Substances that Deplete the Ozone Layer was signed. The Protocol set limits for production and consumption of the damaging CFCs and halons, so it will curb the levels of chlorine and bromine reaching the stratosphere and damaging the ozone layer (see box opposite). The Montreal Protocol entered into force on 1 January 1989.

In the same year, the Parties to the Protocol established four review panels to prepare assessments on various aspects of the ozone problem (scientific assessment, environmental effects, technological

Adjustments to the Montreal Protocol (London 1990)

CFC 11, CFC 12, CFC 113, CFC 114 and CFC 115

- From 1 July 1991 to 31 December 1992, and thereafter, the annual level of consumption and production should not exceed 150 per cent of the 1986 level.
- As of 1 January 1995 and within 12 months, and thereafter, the annual level of consumption and production should not exceed 50 per cent of the 1986 level.
- As of 1 January 1997 and within 12 months, and thereafter, the annual level of consumption and production should not exceed 15 per cent of the 1986 level.
- As of I January 2000 and within 12 months, and thereafter, the consumption and production should be zero.

Halons 1301, 1211 and 2402

- As of 1 January 1992 and within 12 months, and thereafter, the annual level of consumption and production should not exceed the 1986 level.
- As of 1 July 1995 and within 12 months, and thereafter, the annual level of consumption and production should not exceed 50 per cent of the 1986 level.
- As of I January 2000 and within 12 months, and thereafter, consumption and production should be zero.

aspects and economic assessment). The results of these and other studies (6, 19, 22, 23) have shown that the global ozone depletion problem is more imminent and severe than was indicated by the consensus political/scientific view prior to the Montreal negotiations. The studies pointed out that it is highly desirable to phase out CFCs completely by 2000. At their second meeting in June 1990 in London, the Parties to the Protocol agreed to phase out CFCs and halons by 2000 and set a timetable to phase out other compounds (see boxes on page 19 and below). A Multilateral Fund involving UNEP, UNDP and the World Bank was established. Contributions to the Fund are provided by industrialized countries that are Parties to the Protocol and by those developing countries with a per capita consumption of more than 0.3 kg of the controlled substances per year. The Fund is to help developing countries meet the costs of complying with the revised Montreal Protocol, and to provide for the necessary transfer of technology. The Parties also agreed on a mechanism for decision making regarding the Fund in which developed and developing countries have equal representation.

Measures have already been taken in some countries to reduce or ban the use of the controlled CFCs in all or some products (such as non-essential aerosols). The United States took such restrictive

Amendments to the Montreal Protocol (London 1990)

CFC 13, 111, 112, and 211 to 217

- As of 1 January 1993 and within 12 months, and thereafter, the annual level of consumption and production should not exceed 80 per cent of the 1989 level.
- As of 1 July 1997 and within 12 months, and thereafter, the annual level of consumption and production should not exceed 15 per cent of the 1989 level.
- As of 1 July 2000 and within 12 months, and thereafter, consumption and production should be zero.

Carbon tetrachloride

- As of 1 January 1995 and within 12 months, and thereafter, the annual level of consumption and production should not exceed 15 per cent of the 1989 level.
- As of 1 January 2000 and within 12 months, and thereafter, consumption and production should be zero.

Methyl chloroform

- Phase out production and consumption by 2005, with intermediate cuts of 30 per cent by 1995 and 70 per cent by 2000 of the 1989 level.

All CFC substitutes (HCFC 21, 22, 31, 121–124, 131-133,141,142,151, 221–226, 231–235, 241–244, 251–253, 261, 262, 271) have been included on a separate list with a requirement for annual reports on their production and consumption, strict guidelines for their use plus a commitment to phase them out within a specified period. The replacement HCFCs have lower atmospheric lifetimes and lower chlorine-loading potentials than the fully halogenated CFCs and are therefore less ozone-depleting. However, they are considered as 'bridging' chemicals that should be phased out by 2020–2040. Completely acceptable substitutes for long-term use must have no ozone-depleting or global-warming potential.

measures long before the adoption of the Montreal Protocol. Canada, Sweden, Norway, Switzerland and Belgium have banned or drastically restricted the use of CFCs in non-essential aerosols. Several governments have followed suite, and some (such as Germany and the Nordic countries) are now advocating much higher targets for reducing production and use of ozone-depleting substances over the next few years. At the London meeting, Australia, Austria, Belgium, Canada, Denmark, Finland, the Federal Republic of Germany, Liechtenstein, The Netherlands, New Zealand, Norway, Sweden and Switzerland declared their firm determination to take all appropriate measures to phase out the production and consumption of all fully halogenated chlorofluorocarbons controlled by the Montreal Protocol, as adjusted and amended, as soon as possible but not later than 1997.

Industries and financial institutions are also working to reduce the production and use of CFCs and halons. The major chemical industries have announced policies to phase out production of CFCs as soon as safer alternatives are available. Some have set a goal of 1995 for halting CFC production, others will phase out production by 2000. These phase-out policies send customers a strong message to seek alternatives and substitutes. Some industry associations and individual companies have already phased out the use of controlled ozone-depleting substances. Many industry associations are engaged in extensive education, training and public awareness programmes (especially through such measures as the voluntary labelling of products as 'ozone-friendly'). In addition, the major chemical manufacturers of CFCs and halons have pledged not to sell or license CFC-or halon-manufacturing technology to countries that are not parties to the Montreal Protocol.

The major chemical industries have announced policies to phase out production of CFCs as soon as safer alternatives are available. Some have set a goal of 1995 for halting CFC production, others will phase out production by 2000.

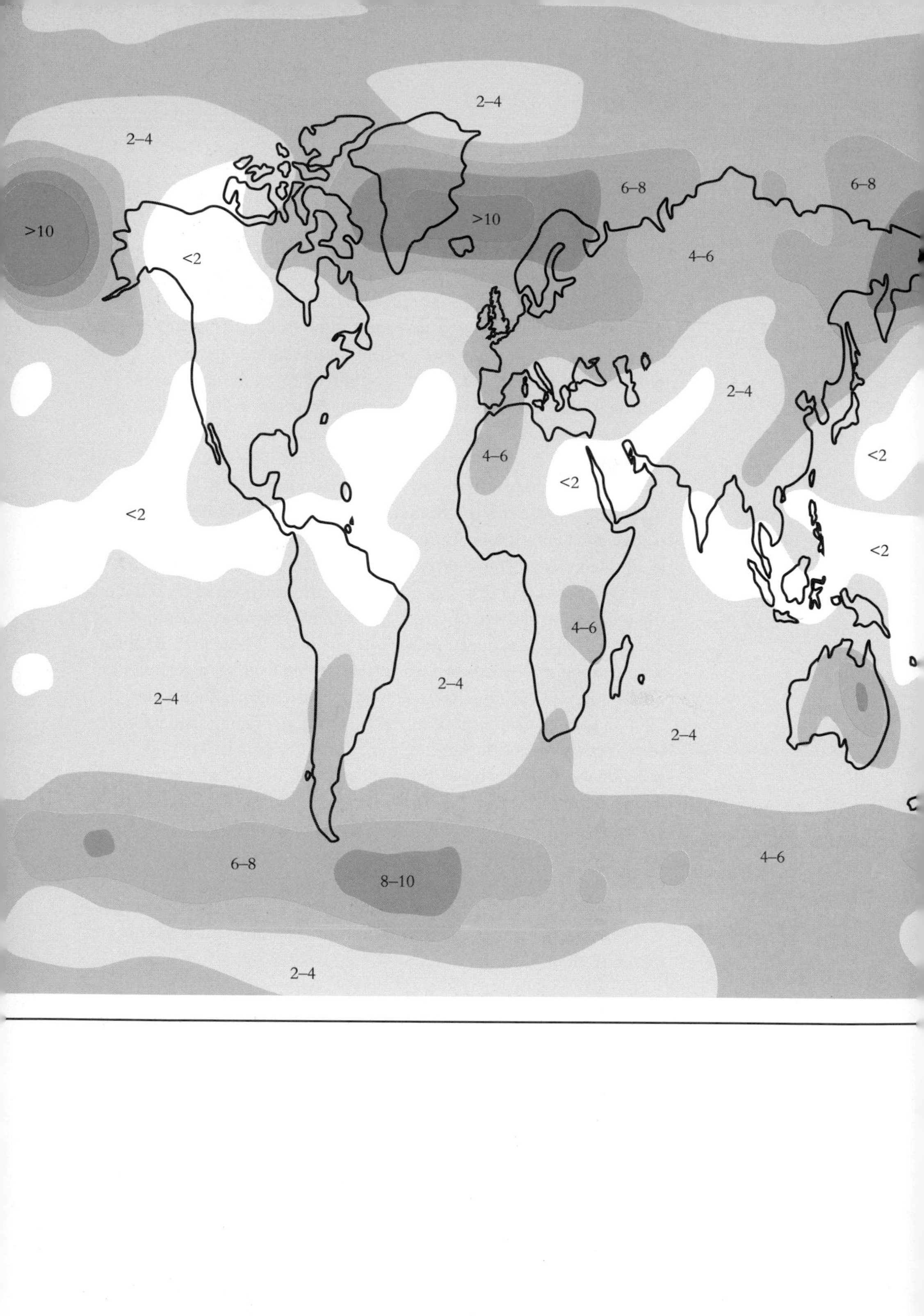

2–4
2–4
6–8
6–8
>10
>10
<2
4–6
2–4
<2
4–6
<2
<2
<2
4–6
2–4
2–4
2–4
6–8
8–10
4–6
2–4

Chapter 3

Climate change

Weather (the day-to-day fluctuations of the atmosphere) and climate (normally taken as a moving 30-year average of weather) are important determinants of a region's energy use, its growth of vegetation, its means of transportation, its water supplies and its patterns of habitation and development. Events such as several years of dry or wet conditions constitute a variation in climate. Climate change, on the other hand, refers to shifts in normal climate—generally in the same direction—lasting over decades.

A primary descriptor of climate is temperature. Sunlight heats up the sea and land. The warmed surface of the earth then radiates heat back towards space. On its way out some of this heat (infrared radiation) is absorbed by trace gases—notably carbon dioxide and water vapour—in the atmosphere and, thereby, keeps the earth's temperature suitable for life. Without this natural greenhouse effect of carbon dioxide and water vapour, the temperature at the earth's surface would be some 33 °C lower than it is today—well below freezing point. The natural concentration of carbon dioxide in the atmosphere is controlled by the interactions of the atmosphere, oceans and the biosphere in what is known as the geochemical carbon cycle. Human activities can disturb this cycle by injecting carbon dioxide into the atmosphere. This leads to a net increase in carbon dioxide concentration in the atmosphere which 'enhances' the natural greenhouse effect.

Although the greenhouse effect has been known for more than a century, it was not until the late 1960s that concern was voiced about the implications of global warming. Studies published in the early 1970s (1, 2) warned of the long-term potential consequences of carbon dioxide accumulation in the atmosphere for the climate. The World Climate Conference convened in 1979 (3) pointed out that some effects of climate change on a regional and global scale may be detectable before the end of this century and may become significant before the middle of the next century. The extensive studies carried out in the 1980s contributed a great deal to our understanding of the problem of climate change.

The greenhouse gases

It was once thought that carbon dioxide was the only greenhouse gas. Research over the past two decades has, however, identified other gases such as nitrous oxide, methane, chlorofluorocarbons and tropospheric ozone as potential greenhouse gases.

The atmospheric carbon dioxide concentration is now 353

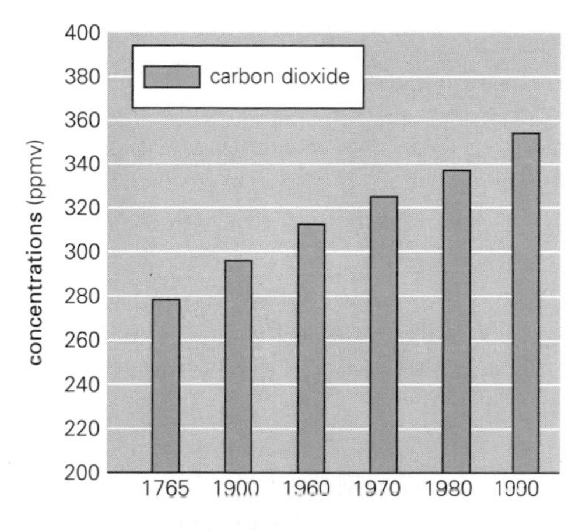

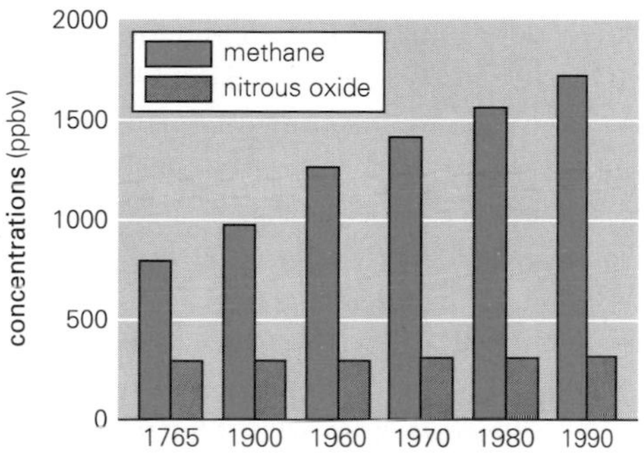

Figure 3.1
Increase in concentrations of carbon dioxide (top) and of methane and nitrous oxide (immediately above) since 1765

based on data from (5)

ppmv, 25 per cent greater than the pre-industrial (1750–1800) value of about 280 ppmv (4, 5) and is currently rising at a rate of about 0.5 per cent per year due to anthropogenic emissions (Figure 3.1). The latter are estimated to be about 5700 million tonnes of carbon per year due to fossil fuel burning, plus 600 to 2500 million tonnes of carbon per year due to deforestation (5, 6). About 40–60 per cent of the carbon dioxide emitted into the atmosphere remains there, at least in the short term; the rest is taken up by natural sinks, particularly the ocean. Future atmospheric carbon dioxide concentrations depend on the amounts of carbon dioxide released from future fossil fuel burning (which are determined by the amount and type of energy sources to be used), the carbon dioxide released from biotic sources (which is determined by the rate of future deforestation and changes of other vegetative cover), and the uptake of carbon dioxide by various natural sinks. The Intergovernmental Panel on Climate Change (IPCC) has estimated that if anthropogenic emissions of carbon dioxide could be kept at present day rates, atmospheric carbon dioxide would increase to 460–560 ppmv by the year 2100 (5) because of the long residence time of carbon dioxide in the atmosphere and the long lead time for its removal by natural sinks.

The current atmospheric methane concentration is 1.72 ppmv, more than double the pre-industrial value of about 0.8 ppmv, and is increasing at a rate of about 0.9 per cent per year. Methane is produced by anaerobic bacteria in natural wetland ecosystems, but the bulk of methane is produced through human activities such as rice cultivation, the rearing of domestic ruminants, biomass burning, coal mining and natural gas venting. The total annual flux to the atmosphere is between 400 and 600 million tonnes of methane a year (5, 7, 8,

The atmospheric carbon dioxide concentration is now 353 ppmv, 25 per cent greater than the pre-industrial (1750–1800) value of about 280 ppmv and is currently rising at a rate of about 0.5 per cent per year due to anthropogenic emissions.

9). Of this amount, natural wetland ecosystems account for 100–150 million tonnes; rice paddies contribute an average of about 110 million tonnes of methane per year (5). Recently, a similar figure of 100 million tonnes of methane per year has been given (10); about half this amount is due to rice cultivation in China alone.

The mean atmospheric concentration of nitrous oxide in 1990 was about 310 ppbv, about 8 per cent greater than the pre-industrial value of about 285 ppbv (5). Nitrous oxide emissions result naturally from microbial processes in soil and water (about 4.3 to 7.8 million tonnes of nitrogen per year). Human activities add to them about 0.1 to 2.7 million tonnes of nitrogen per year from burning biomass and fossil fuels. The atmospheric nitrous oxide concentration is increasing at a rate of about 0.2–0.3 per cent per year.

Most halocarbons, with the exception of methyl chloride, are exclusively of industrial origin. The atmospheric concentration of methyl chloride is about 0.6 ppbv, and is primarily released from the oceans. The atmospheric concentrations of the other halocarbons (especially CFC 11, 12 and 113, and methyl chloroform) have increased rapidly over the past few decades (see Chapter 2). Future emissions of most halocarbons will be greatly reduced by the year 2000, according to the amended and adjusted Montreal Protocol (Chapter 2). However, the atmospheric concentrations of CFCs 11, 12 and 113 will still be significant (about 30–40 per cent of current concentrations) for at least the next century because of their long atmospheric lifetimes.

The contribution of the these trace gases to the greenhouse effect depends on the amount of the gas released into the atmosphere, its net concentration in the atmosphere, its lifetime, and its radiative

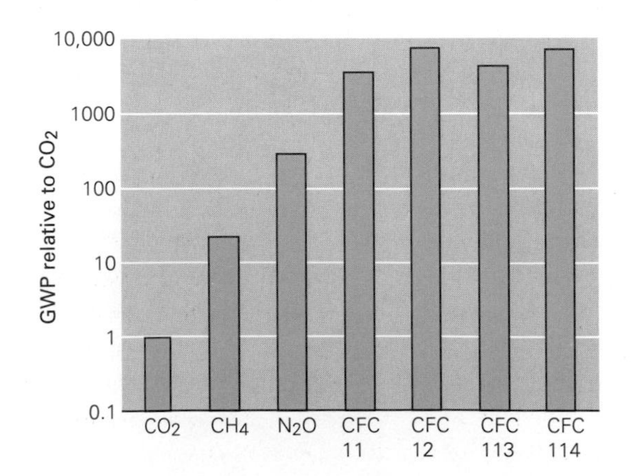

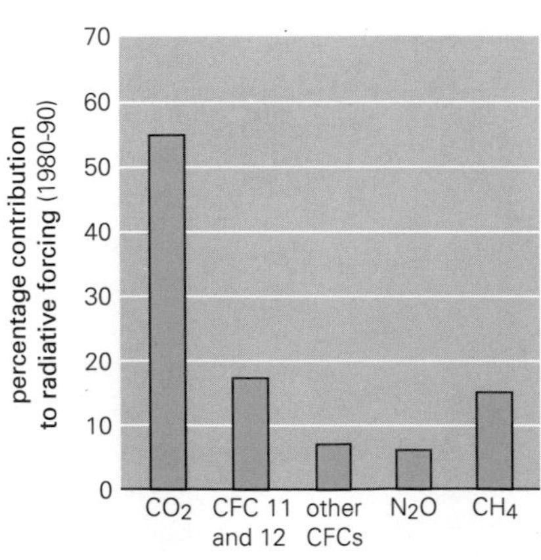

Figure 3.2
Global warming potential of greenhouse gases* (below) and contribution of greenhouse gases to radiative forcing, 1980–90 (below right)

* integrated 100 years time horizon

based on data from (5)

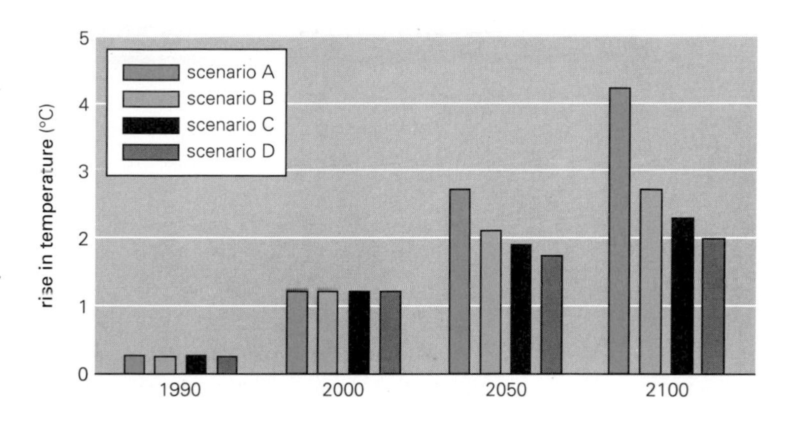

Figure 3.3
Projected rise in temperature above pre-industrial (1765) times

Scenarios as defined by IPCC:

Scenario A
Energy supply and demand continue as they are. Deforestation continues at present rate. Partial implementation of Montreal Protocol (business-as-usual).

Scenario B
Energy supply mix shifts towards low carbon fuels and natural gas. More energy efficiency. Deforestation reversed. Full implementation of Montreal Protocol.

Scenario C
Shift towards renewable sources of energy and nuclear power in second half of next century.

Scenario D
Shift towards renewable sources of energy and nuclear power in first half of next century.

based on data from (5)

forcing. The global warming potential (GWP)—defined as the time-integrated warming effect due to a release of 1 kg of a given greenhouse gas in today's atmosphere, relative to that of carbon dioxide—shows that carbon dioxide is the least effective greenhouse gas per kilogram emitted (Figure 3.2). However, its contribution to global warming, which depends on the product of the GWP and the amount emitted, is the largest.

Scenarios of climate change

The task of predicting future climate change is extremely complex. The effects of the build-up of greenhouse gases in the atmosphere cannot be studied directly. Over the past two decades, a hierarchy of climate models (mathematical representations of the atmosphere used to simulate climate change under different scenarios) has been developed to estimate climate change. Early estimates made during the late 1960s predicted that a doubled CO_2 concentration in the atmosphere should raise the average temperature by 1.5–3.0 °C (11) . More than 100 independent estimates of average surface temperature increase have since been made. Almost all lie in the range 1.5–4.5 °C, with values near 3.0 °C tending to be favoured (12, 13). The IPCC has recently predicted that under the 'business-as-usual' scenario (without actions to reduce emissions of greenhouse gases) global warming could reach 2–5 °C over the next century (with the best estimate around 3 °C), a rate of change unprecedented in the past 10 000 years (5). The rate of increase of global mean temperature during the next century would be about 0.3 °C per decade (Figure 3.3). Uncertainties in predictions of climate change revolve around the timing, regional patterns and impacts of climate change.

Has the climate actually changed?

Over the past 100 years, the atmospheric CO_2 concentration has increased by about 25 per cent. A range of model calculations suggests that the corresponding equilibrium temperature rise should be 0.5–1.0 °C. If this is corrected for the effects of the thermal inertia of the oceans (which delays climate change for a period of 10–20 years), the changing composition of the atmosphere should by now have produced a warming of 0.35–0.7 °C superimposed on the natural fluctuations of the atmosphere (14).

Detailed analysis of temperature records of the past 100 years indicates that the global mean temperature has risen by 0.3-0.6 °C (Figure 3.4). Much of the warming since 1900 has been concentrated in two periods, the first between about 1910 and 1940, and the second since 1975; the five warmest years on record have all been in the 1980s (5, 15). The size of the warming over the past century is broadly consistent with the predictions of climate models, but is also of the same magnitude as natural climate variability.

Impacts of climate change

Sufficient evidence is now available to indicate that changes in climate would have an important effect on agriculture and livestock. Negative impacts could be felt at the regional level as a result of changes in weather and the arrival of pests associated with climate change, necessitating innovations in technology and agricultural management practices. There may be a severe decline in production in some regions (such as Brazil, the Sahel region of Africa, South-East Asia and the Asian region of the Soviet Union and China), but there may be an

Figure 3.4
Global change in temperature (1861–1989)

source: (5)

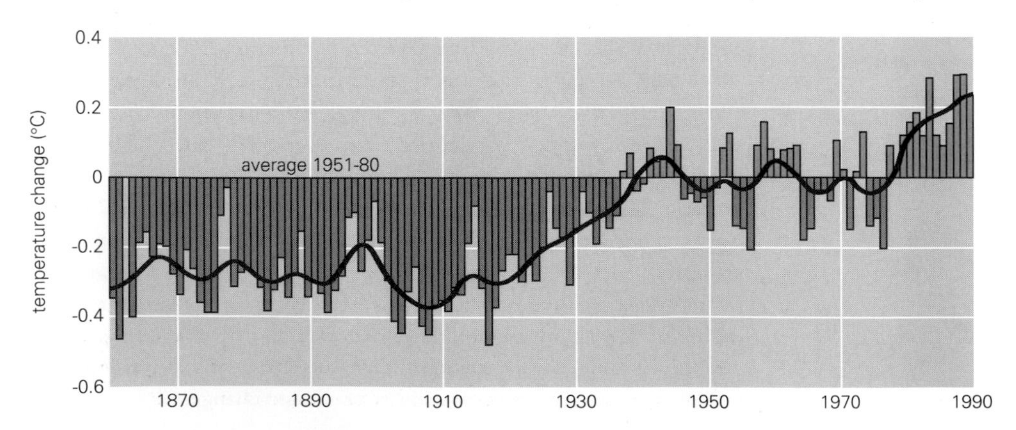

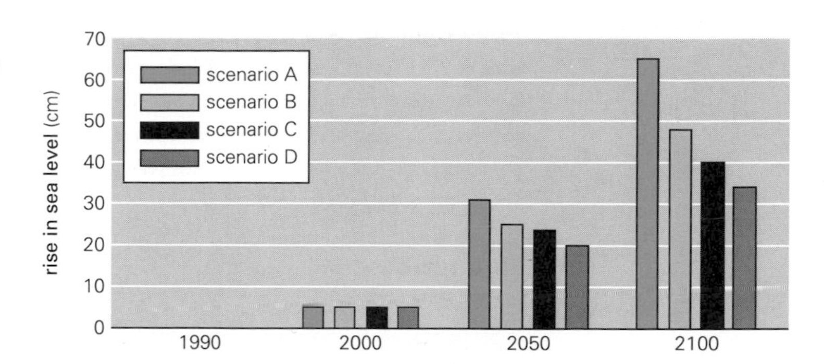

Figure 3.5
Projected sea level rise

for definition of scenarios, see Figure 3.3

based on data from (5)

increase in production in other regions because of a prolonged growing season. The effects of global warming on forests may also be mixed, and will vary from one region to another (5).

Terrestrial ecosystems could face significant consequences. Projected changes in temperature and precipitation suggest that climatic zones could shift several hundred kilometres towards the poles over the next 50–100 years. Flora and fauna would lag behind these climatic shifts, surviving in their present location and, therefore, could find themselves in a different climatic regime. These regimes may be more or less hospitable and, therefore, could increase the productivity of some species and decrease that of others. Ecosystems are not expected to move as a single unit, but would develop new structures as the distribution and abundance of species changed.

Relatively small climate changes can cause large water resource problems in many areas, especially in semi-arid regions and those humid areas where demand or pollution has led to water scarcity. Little is known about the regional details of hydrometeorological change induced by greenhouse warming. Many areas may have increased precipitation, soil moisture and water storage, thus altering patterns of agricultural, ecosystem and other water use. Water availability will decrease in other areas, with grave implications for water-critical areas such as the Sahelian zone in Africa.

Global warming will accelerate sea-level rise (Figure 3.5), modify ocean circulation and change marine ecosystems, with considerable socioeconomic consequences. The IPCC predicted that under the 'business-as-usual' scenario an average rate of global mean sea level rise of about 6 cm per decade could occur over the next century. The predicted rise is about 20 cm in global mean sea level by 2030, and 65 cm by the end of the next century, and there will be significant regional variations (5). A sea level rise of this magnitude will threaten low-lying islands and coastal zones. It will render some island countries uninhabitable, displace tens of millions of people, seriously threaten low-lying urban areas, flood productive lands, contaminate freshwater supplies and change coastlines. In coastal lowlands such as in Bangladesh, China and Egypt inundation due to sea level rise and storm surges could lead to significant social disruptions and economic losses.

Responses

The prospective global warming and the forces driving it are now broadly understood. A clear scientific consensus has emerged on estimates of the range of global warming that can be expected during the 21st century, notwithstanding uncertainties about its precise regional distribution and its environmental consequences. Based on current scientific findings, the world community has two options.

The first option is to consider the issue academic and to let things go on as at present. If this happens,the world will eventually have to adapt its socioeconomic structure suddenly to the changing climate, and face possibly catastrophic consequences. This is clearly unviable.

The second option is to apply the anticipatory principle and take immediate measures to slow down the build-up of greenhouse gases in the atmosphere, and hence minimize the warming and its potential undesirable consequences. The Second World Climate Conference (1990) concluded that 'nations should now take steps towards reducing sources and increasing sinks of greenhouse gases through national and regional actions, and negotiation of a global convention on climate change and related legal instruments. The long-term goal should be to halt the build-up of greenhouse gases at a level that minimizes risks to society and natural ecosystems'.

The Montreal Protocol on Substances that Deplete the Ozone Layer (Chapter 2) is a step in the right direction. It calls for a complete phase-out of the main halocarbons by the year 2000. But it is insufficient to deal with the problem of global warming. Carbon dioxide and methane account for 70 per cent of the increased radiative forcing produced by greenhouse gases from human activities, and the focus should be on strategies and tools that would freeze or reduce the rate of their emission into the atmosphere.

The Second World Climate Conference (1990) concluded that 'technically feasible and cost-effective opportunities exist to reduce carbon dioxide emissions in all countries'. Increasing the efficiency of energy use, and employing environmentally-sound alternative energy sources, especially renewable ones, will contribute significantly to the reduction of CO_2 emissions. In addition, reversing the current net losses of forests would increase storage of carbon. There are already encouraging signs of a concerted global effort to control emissions of greenhouse gases. Several developed countries have committed themselves to stabilizing their year 2000 emissions of CO_2, and other greenhouse gases not controlled by the Montreal Protocol, at 1990 levels. The countries involved include the European Community with its member states, Australia, Austria, Canada, Finland, Iceland, Japan, New Zealand, Norway, Sweden and Switzerland.

Although a number of international legal mechanisms exist that have a bearing on the climate change issue, they are insufficient to meet the challenge. An international consensus emerged at the 44th session of the United Nations General Assembly on the need to prepare, as a matter of urgency, a framework convention on climate change with specific commitments. This consensus was reiterated by the Ministerial Declaration of the Second World Climate Conference in 1990. The IPCC suggested that a framework convention should articulate a multilateral greenhouse gases control strategy, while simultaneously encouraging unilateral action by the largest emitters and the establishment of specific national commitments. A global climate convention should establish global goals regarding future emissions of greenhouse gases. This agreement should also address other institutional issues, such as cooperation with developing countries in the areas of additional financial resources and transfer of technology as well as in the establishment of efficacious decision-making processes. Protocols to establish specific national requirements to assure attainment of global targets set out in the convention should be negotiated simultaneously with the convention (5, 16). Rounds of negotiation are under way to draft such global climate convention, and it is hoped that the convention will be ready for signature at the time of the United Nations Conference on Environment and Development to be convened in Brazil in June 1992.

There are encouraging signs of a concerted global effort to control emissions of greenhouse gases. Several developed countries have committed themselves to stabilizing their year 2000 emissions of CO_2 at 1990 levels.

Chapter 4

Marine pollution

The oceans cover 71 per cent of the Earth's surface and—through their interactions with the atmosphere, lithosphere and biosphere in what is known as the geochemical cycles—have played a major role in shaping the conditions that have made life possible on Earth. In addition to serving as the habitat for a vast array of plants and animals, the oceans also supply people with food, energy and mineral resources. More than half the people in developing countries obtain 30 per cent or more of their animal protein from marine fish (Chapter 11).

For geological ages, the oceans have received natural dissolved and suspended matter, especially from continents. Rivers have delivered to the oceans annually about 35 trillion tonnes of water, 3.9 billion tonnes of dissolved matter and between 10 and 65 billion tonnes of suspended particulate matter (1). Additional inputs come from groundwater discharging through the continental shelf, submarine springs of volcanic and deeper crustal origin, and the atmosphere through which airborne gases and particulates reach the oceans. The volume and composition of the oceans have—remarkably—been held more or less unchanged for a geologically long period as a result of the balance established by the geochemical cycles.

Human activities on land and in the sea are disturbing this balance, and changing the composition of seawater. This is most marked along coastal and near-shore areas, which are among the most intensively used parts of the earth. About 60 per cent of the world's population, or nearly 3 billion people, live on or within some 100 km of the shoreline. The coastal zones are sites for large-scale industrial development, are intensively used for recreation, and include the world's harbours on which international trade depends. Coastal areas contain many ecosystems that are vital to marine life and humankind;

Figure 4.1
Discharges in the marine environment (late 1980s)

based on (5, 16)

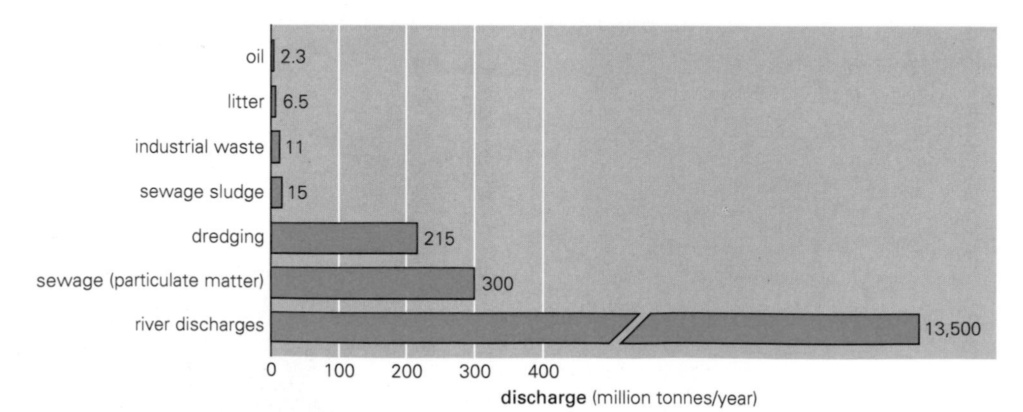

four of the most productive are salt marshes, mangroves, estuaries and coral reefs. About 95 per cent of the world fish catch comes from near-shore areas.

The open oceans still seem to be largely unaffected by human activities. However, the marine environment in coastal areas, and in enclosed and semi-enclosed seas, has generally deteriorated over the past two decades (in spite of local improvements here and there). The symptoms of the malaise include the spread of algal blooms, coral bleaching, epidemics, oil pollution, and a decline in the quality and quantity of marine food.

Sources of marine pollution

The two dominant pathways by which potential pollutants reach the oceans from the continents are the atmosphere and rivers. The atmospheric pathway accounts for more than 90 per cent of the lead, cadmium, copper, iron, zinc, arsenic, nickel, PCBs, DDT and HCH found in the open oceans (2). River inputs are generally more important than those from the atmosphere in coastal zones, although in certain areas and for some substances (such as lead and HCH in the North Sea, and nitrogen in the Mediterranean) atmospheric inputs are as important or even dominant.

Aside from physical degradation, pollution is the major problem affecting the coastal and near-shore zones. Most of the liquid wastes and a growing fraction of solid wastes resulting from man's activities on land are introduced into the oceans through the land/sea interface. Coastal areas receive direct discharges from rivers, surface run-off and drainage from the hinterland, domestic and industrial effluents through outfalls, and various contaminants from ships (Figure 4.1).

Some 6.5 million tonnes of litter finds its way into the sea each year. In the past, much of such solid matter disintegrated quickly, but resistant synthetic substances have been replacing many natural, more degradable materials. Plastics, for example, persist for up to 50 years and, because they are usually buoyant, they are widely distributed by ocean currents and the wind. Most beaches near population centres are littered with plastic residues washed up from the sea, contributed by rivers, ships and outfalls, dumped by illegal refuse

The marine environment in coastal areas ... has generally deteriorated over the past two decades ... The symptoms of the malaise include the spread of algal blooms, coral bleaching, epidemics, oil pollution, and a decline in the quality and quantity of marine food.

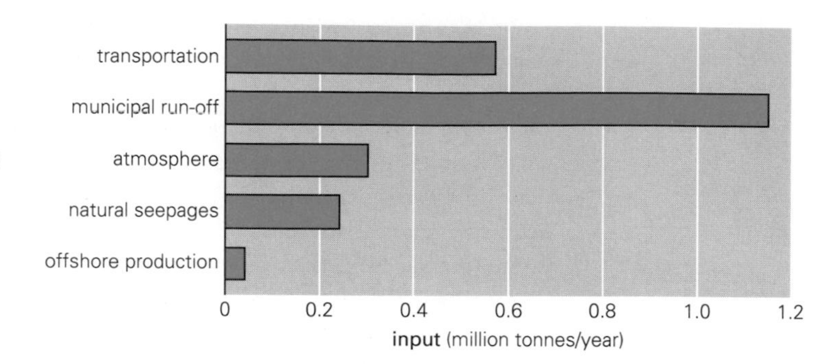

Figure 4.2
Input of petroleum to the marine environment

based on data from (4, 5)

operators, or left behind by beach users. A major source of plastic debris is the fishing industry; it has been estimated that more than 150 000 tonnes of fishing gear made of plastic is lost (or discarded) in the oceans each year (2). In one study, up to 70 percent of the debris examined along the beaches of the Mediterranean was plastic; in the Pacific, the figure reached more than 80 percent (3).

In 1985, the National Research Council (4) estimated that the amount of petroleum, from all sources, entering the marine environment was 22.3 million barrels (3.2 million tonnes) annually. Of this amount, municipal wastes and run-off accounted for 8.1 million barrels (1.16 million tonnes), and maritime transportation for about 10.1 million barrels (1.47 million tonnes). However, recent data (5) indicate that the amount of oil entering the world's oceans as a result of shipping operations has been cut by 60 per cent since 1981. In 1989, it was estimated that maritime transportation accounted for about 4 million barrels (568 800 tonnes) of spilt oil annually, of which tanker accidents were responsible for about 20 per cent (Chapter 9). Figure 4.2 gives estimates of the quantities of oil entering the marine environment. A major reason for the improvement in the pollution figures has been the entry into force of the International Convention for the Prevention of Pollution from Ships, 1973, as modified by the Protocol of 1978 (MARPOL 73/78), in 1983. The Convention applies now to more than 85 per cent of the world fleet of merchant ships.

Impacts of marine pollution

Most types of wastes, once introduced into the sea, cannot be removed from there. Their fate is determined by their chemical composition and by the physical transport processes (such as mixing and sea currents) to which they are subjected. The distance they travel depends on these processes and on their rate of decomposition; non-degradable wastes can travel long distances.

Some wastes are easily decomposed to harmless substances, although their end products, if excessively concentrated, may lead to serious disturbances of ecosystems (as when eutrophication results from excess nutrients). Other wastes, such as metals and persistent organic compounds, cannot be degraded; they usually remain adsorbed to the bottom sediments near the source of discharge. Some marine organisms have a remarkable ability to accumulate such substances from seawater, even when the materials are present in extremely low concentrations. Others can convert some compounds into more toxic ones—the conversion of inorganic mercury into methyl mercury, for example, caused the outbreak of Minamata disease in Japan in the 1950s and 1960s.

Globally, the principal problem for human health from marine pollution is the existence of pathogenic organisms discharged with domestic sewage into coastal waters. Bathing in sewage-polluted seawater, and consumption of contaminated fish and shellfish, are the causes of a variety of infections. Epidemiological studies show that those who swim in sewage-polluted seawater have an above-normal incidence of gastric disorders (2), and an increased incidence of non-gastric disorders such as ear, respiratory and skin infections. The consumption of contaminated seafood is linked with serious illness, including viral hepatitis and cholera.

Both sewage and agricultural run-off introduce large quantities of nitrogen and phosphorus into coastal water. These compounds, from sources such as detergents, fertilizers, and human and animal waste, nourish algae and can cause explosive algal growths which deplete the water of oxygen and suffocate other species. Oxygen-depleted waters are known as 'dead zones'; a 4000 square-kilometre dead zone has been found in the Gulf of Mexico, near the mouth of the Mississippi River.

Algal clusters can block sunlight and stunt the growth of other marine life. Over the past two decades, the frequency of algal blooms has been increasing in coastal areas. Some of the algae produce toxins which are detrimental, even fatal, to other marine life. The toxins may also be consumed by other organisms, become enriched in the marine food chain, and

Oxygen-depleted waters are known as 'dead zones'; a 4000 square-kilometre dead zone has been found in the Gulf of Mexico, near the mouth of the Mississippi River.

ultimately affect people who consume marine food. An outbreak of paralytic shellfish poisoning (PSP) in Guatemala in 1987 killed 26 people; the organism involved is believed to have been a toxic alga. The incidence of PSP is increasing globally (6).

The term 'red tide' is often used to describe the discolouration of water caused by any algal bloom. Red tides (dominantly toxic) are annual events in many parts of the world. Japan's Inland Sea is affected by some 200 red tides each year. The number of red tides in Hong Kong Harbour increased from two in 1977 to 19 in 1987 (7). Blooms of toxic species occurred in the North Sea with increasing frequency in the 1970s and 1980s (8). In 1988, a massive bloom occurred in the seas around southern Scandinavia, damaging marine life in some seas and some fish farms along the coast of Norway (8). Although unusual occurrences of algal blooms have been attributed to a combination of many factors, especially to disturbances in the marine ecological balance caused by climatic factors, considerable evidence suggests that the increased incidence of blooms is related to the nutrient enrichment of coastal waters and inland seas on a global scale.

Many compounds discharged into the sea tend to accumulate in marine organisms. Halogenated hydrocarbons accumulate in fatty tissues, and the amount accumulated may increase along the food chain, so that high concentrations are found in the bodies of the top predators among birds, fish and mammals. Where the contamination has built up over decades, as in enclosed areas such as the Baltic and the Wadden Sea, the reproductive capacity of marine mammals and birds has been affected (2). Polychlorinated biphenyls (PCBs) accumulated in seafood can reach levels that render the food unacceptable for the market. Tributyltin (TBT) affects a wide range of invertebrates and its use in marine paints was recently restricted in France, the United Kingdom, and several states in the United States.

Oil in the sea is normally found in concentrations too low to pose a threat to marine organisms. However, oil spills, especially major ones, may cause excessive damage, especially in coastal areas (Chapter 9).

Several human activities have direct effects on coastal areas, especially on sensitive ecosystems such as salt marshes, mangroves and coral reefs. For example, mangrove forests on the East African coast have been depleted for fuelwood and building materials. Along East Asian coasts, extensive conversion of mangrove forest to rice fields and fish ponds has eliminated natural barriers to flooding from storms. In 1980 the Philippines had 146 000 ha of mangrove forests; today it has a mere 38 000 ha (9). In Central and South America, mangroves are being cleared for fish farming. Coral reefs also face a variety of threats and are being damaged in some tropical countries by uncontrolled tourism and near-shore human activities. In the Philippines, only 10 per cent of the country's coral reefs are in good condition—the rest has been damaged to varying degrees (10). Many of the world's coastal wetlands have diminished over the past decades as a result of drainage and reclamation schemes to increase land for agriculture, industry and urban growth. Many of the valuable habitats of these wetlands have been lost as a result of these activities (Chapter 8).

Over-exploitation of living marine resources

The world marine fish catch (including aquatic plants) rose from 60 million tonnes in 1970 to 91 million tonnes in 1989 (11, 12). The FAO estimates that the world catch ought not to exceed 100 million tonnes per year if the risk of a substantial depletion of fish stocks is to be avoided. However, pressures on stocks in certain areas already amount to overfishing. Overfishing has led to a sharp drop in catches of cod and herring, and fishing for these species in the north-east Atlantic was made subject to quotas in the 1970s, and subsequently banned altogether for certain stocks, in order to allow stocks to recuperate.

Excessive harvesting of whales, dolphins, seals and polar bears is one of the clearest examples of over-exploitation of marine resources. At its peak,the whaling industry killed some 66 000 whales a year and depleted many species to near extinction. In 1989, new provisional International Whaling Commission figures indicated that of the million sperm whales that once roamed the oceans, only 10 000 are thought to be left. Humpbacks seem to be down from 20 000 to 4000, fin whales from more than 100 000 to 2000, and blue whales from 250 000 to around 500. In 1985, the International Whaling Commission imposed a five-year moratorium on commercial whaling. But since then more than 11 000 whales have been killed (9).

... of the million sperm whales that once roamed the oceans, only 10 000 are thought to be left. Humpbacks seem to be down from 20 000 to 4000, fin whales from more than 10 000 to 2000, and blue whales from 250 000 to around 500.

Responses

Several measures have been taken to control marine pollution. They range from isolated national actions to control pollution from easily identifiable sources in specific sites, to measures to curb pollution on regional levels, and to global approaches to controlling pollution through international agreements.

Historically, international marine agreements dealt with the regulation of navigational fishing. Only recently has it been recognized that the world oceans should be regulated and protected as a natural resource. This important change from a 'user-oriented' to a 'resource-oriented' approach became most marked in the past two decades. Most legal regimes adopted since 1970 encompassed the protection, conservation and management of the marine and coastal environment and their resources. The most significant are: Convention on Wetlands of International Importance, Ramsar, 1971; Convention on the Prevention of Marine Pollution by Dumping of Wastes and Other Matter, London, 1972; International Convention for the Prevention of Pollution from Ships, London, 1973; United Nations Convention on the Law of the Sea; and the several regional seas conventions. UNEP's 1991 register of international legislations lists all those related to the marine environment.

Although the importance of reducing maritime sources of ocean pollution had led to action in the 1960s, it was not until the early 1970s that land-based activities were recognized as the most significant source of marine pollution. The Convention on the Protection of the Marine Environment of the Baltic Sea Area (Helsinki, 1974), and the Convention on the Prevention of Marine Pollution from Land-based Sources (Paris, 1974) were among the first Conventions formulated to control pollution from land-based sources.

Under the catalytic and coordinating role of UNEP, the 'regional seas' programme was begun in the mid-1970s. The Mediterranean states agreed in 1975 on an Action Plan for the Protection of the Mediterranean Environment (MAP). In the following year, the Barcelona Convention for the Protection of the Mediterranean Sea Against Pollution, plus two protocols, were signed. In the same year, a regional oil-combating centre was established in Malta as part of the MAP. In 1979 a 'Blue Plan' for the long-term management of the Mediterranean Sea was launched as part of the socio-economic component of the MAP. It was intended to integrate development plans with environmental protection measures in the Mediterranean Basin. In 1980, the Mediterranean states moved a step forward by adopting the Protocol for the Protection of the Mediterranean Sea Against Pollution from Land-Based Sources. This agreement identifies measures to control coastal pollution from municipal sewage, industrial wastes and agricultural chemicals. Two years later, the Mediterranean governments also approved a protocol providing special protection for endangered species of fauna and flora as well as critical habitats. In 1985, the Mediterranean countries established ten priority targets for the decade 1985–95.

In addition to the MAP, action plans were adopted for regional seas in eight other areas: Kuwait, the Wider Caribbean, West and Central Africa, East Africa, South-east Pacific, Red Sea and the Gulf of Aden, South Pacific and East Asia . An action plan was recently drafted for the South Asian region, and is under consideration for approval by the concerned governments. Other action plans for the Black Sea and the Atlantic are being developed. All told, the regional seas programme involves some 130 countries, 16 UN agencies and more than 40 other international and regional organizations, all working with UNEP to improve the marine environment and make better use of its resources—see (10) for a description of the regional seas programmes.

The adoption of the United Nations Convention on the Law of the Sea in 1982 (as of 31 December 1990, 160 countries had signed the Convention) set up a comprehensive new legal regime for the sea and oceans and, as far as environmental provisions are concerned, established material rules concerning environmental standards as well as enforcement provisions dealing with the pollution of the marine environment. Although the Law of the Sea has not yet entered into force, the concept of the 200-mile exclusive economic zone is already effectively in operation, and the Law may play a key role in the management of the ocean's resources.

The physical and biotic features of Antarctica represent, for the most part, extreme conditions—isolation, cold, windiness, extensive glacial ice and sea ice, impoverished terrestrial biota and abundant marine biota. Antarctica was untouched by man until the past two centuries and is still almost pristine. The Antarctic Treaty, Washington, 1959, seeks to ensure, among other things, that the continent is used for international cooperation in scientific research. The Treaty

The regional seas programme involves some 130 countries, 16 UN agencies and more than 40 other international and regional organizations, all working with UNEP to improve the marine environment and make better use of its resources.

bans military activity, nuclear explosions and the disposal of radioactive waste in the region. In the 1960s and 1970s, the Parties to the Treaty agreed on measures for the conservation of fauna and flora, seals and marine living resources of the region. There is growing concern, however, among several governments and non-governmental organizations that the Antarctic Treaty may not be proving effective in protecting the Antarctic environment, and that direct human activities such as intensive exploration, research and exploitation of living and mineral resources in the region will have both direct and indirect impacts (13, 14, 15). Recently, the Parties to the Antarctic Treaty reached an agreement on a 50-year prohibition of mining and mineral prospecting in the region.

In spite of these efforts to protect the marine environment, progress has been rather slow, especially in the developing regions. The capabilities of most developing countries are still generally insufficient to cope adequately with the assessment of the problems facing their marine and coastal environments, and the rational management of their resources. Weak institutional structures hamper the effective participation of many countries in international efforts designed to protect and develop the marine and coastal environment. The effectiveness of regional agreements to enable a rapid response to accidents involving ships and to combat ensuing environmental threats are constrained by available resources and capabilities. For countries that lack the necessary material resources and trained human resources, the agreements are consequently of limited use. Initiatives to improve this situation include the recently adopted International Convention on Oil Pollution Preparedness Response and Cooperation (OPRC Convention, 1990) which contains a mandatory requirement for oil pollution emergency plans.

The 1990 GESAMP report emphasized that strong coordinated national and international action should now be taken to prevent the rapid deterioration of the marine environment. At the national level in particular, the concerted application of measures to reduce discharges into the sea and to manage coastal areas in a rational and environmentally sound way will be essential.

There is growing concern ... that the Antarctic Treaty may not be proving effective in protecting the Antarctic environment, and that ... intensive exploration, research and exploitation of living and mineral resources in the region will have both direct and indirect impacts

Chapter 5

Freshwater resources and water quality

A number of estimates suggest that of all the water around the globe 94 per cent is salt water in the oceans and 6 per cent is fresh. Of the latter, about 27 per cent is in glaciers and 72 per cent is underground. Less than 1 per cent of the world's fresh water is therefore to be found in the atmosphere, in rivers and streams, and in lakes (1). This freshwater supply is continually replenished by precipitation as rain or snow. It has been estimated that the total annual run-off from continents is about 41 000 cubic kilometres (km^3). Of these, 27 000 km^3 return to the sea as flood run-off, and another 5000 km^3 flow into the sea in uninhabited areas. This leaves 9000 km^3 of water readily available for human exploitation worldwide (2, 3). As both the world's population and usable water are unevenly distributed, the local availability of water varies widely. Much of the Middle East and North Africa, parts of Central America and the western United States are already short of water. By the year 2000, water will be scarce in many countries due to increasing demand from agriculture, industry and domestic use.

Demand for water varies greatly in different countries; it depends on population and on the prevailing level and pattern of socio-economic development. Marked differences exist between developed and developing countries. For example, the average per capita domestic use of water in the United States is more than 70 times that in Ghana. Worldwide water use increased dramatically from about 1360 km^3 in 1950 to 4130 km^3 in 1990 (Figure 5.1) and is expected to reach about 5190 km^3 by 2000 (4). Agriculture is the main drain on water supply. Averaged globally, 69 per cent of water

Figure 5.1
Water withdrawal by region (immediately below) and increase in water withdrawal by region, 1970–90 (bottom right)

based on data from (4)

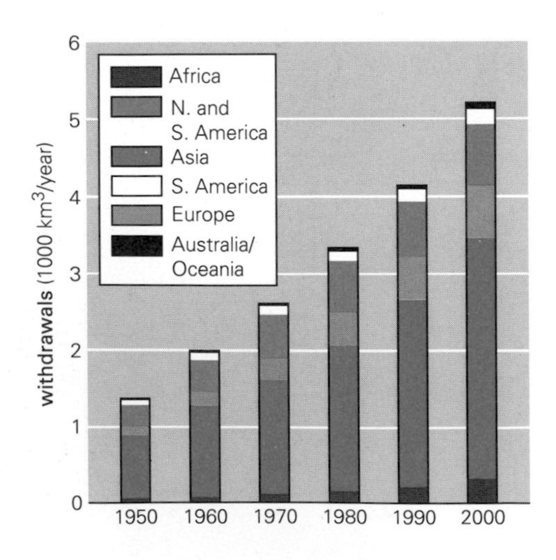

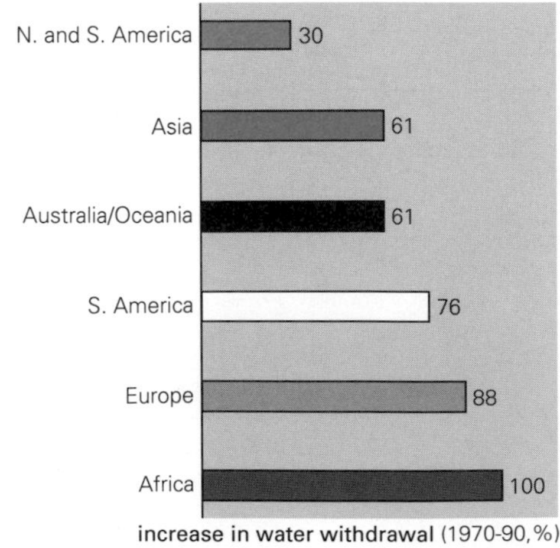

Agriculture is the main drain on water supply. Averaged globally, 69 per cent of water withdrawn is used in agriculture, 23 per cent in industry and 8 per cent for domestic purposes

withdrawn is used in agriculture, 23 per cent in industry and 8 per cent for domestic purposes (Figure 5.2).

Assuring an adequate supply is not the only water problem: countries also need to worry about water quality, concern over which has been growing since the 1960s. At first, attention centred on surface water pollution from point sources, but more recently groundwater, sediment pollution and non point sources have been found to be at least equally serious problems.

Considerable water pollution is caused by the discharge of untreated or inadequately treated wastewater into rivers, lakes and reservoirs. As industry has grown, the discharge of industrial wastewater has created new pollution problems. Another water quality problem is the increasing eutrophication of rivers and lakes caused mainly by fertilizer run-off from agricultural land. Acidification of lakes by acidic deposition is common in some European countries and in North America (Chapter 1). Wastes can also be carried to lakes and streams along indirect pathways—for example, when water leaches through contaminated soils and transports the contaminants to a lake or river. Dumps of toxic chemical waste on land have become a serious source of groundwater and surface water pollution (Chapter 10). In areas of intensive animal farming, or where large amounts of nitrate fertilizers are used, nitrates in groundwater often reach concentrations that exceed guidelines established by WHO. The problem is causing concern in some European countries and in the

Figure 5.2
Water withdrawal by sector (immediately below) and increase in water withdrawal by sector, 1970–90 (bottom right)

based on data from (4)

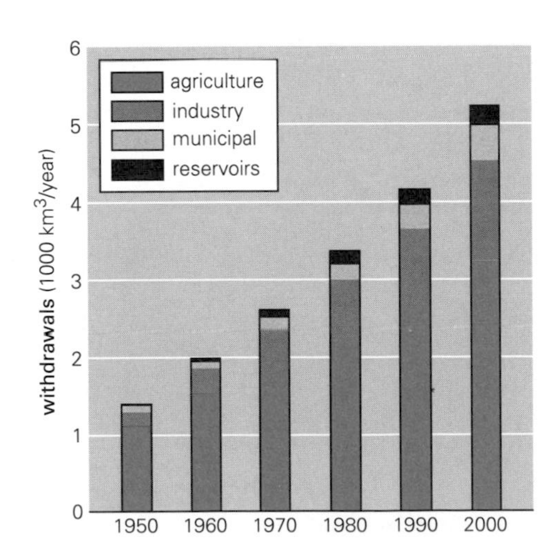

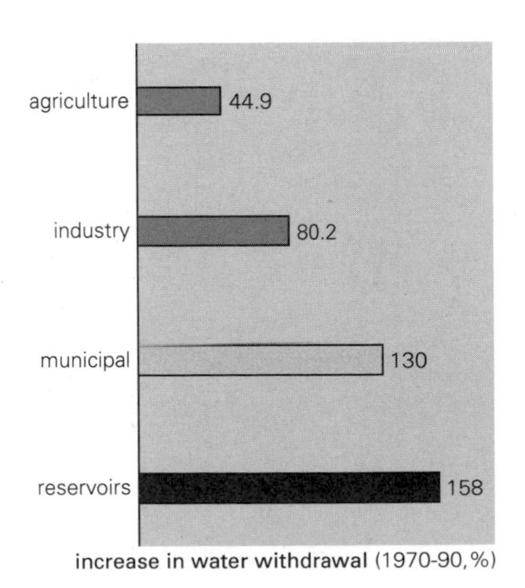

United States, and is growing in magnitude in some developing countries (Chapter 11 and 18).

Water quality monitoring has been introduced in several countries. The GEMS global water quality monitoring project (UNEP/WHO/WMO/UNESCO), launched in 1977, consists of 344 stations (240 river, 43 lake and 61 groundwater stations) in 59 countries. The GEMS/WATER project provides for the collection of data on about 50 different parameters of water quality, including basic measurements such as dissolved oxygen, biochemical oxygen demand (BOD), faecal coliform and nitrates, as well as analyses of chemical trace constituents and contaminants (heavy metals and organic micropollutants).

About 10 per cent of all the rivers monitored in the GEMS/WATER project may be described as polluted, as they have a BOD of more than 6.5 mg/l (5). The two most important nutrients, nitrogen and phosphorus, are well above natural levels in the waters measured by the network. The median nitrate level in unpolluted rivers is 100 micrograms/l (μg/l). The European rivers monitored by GEMS have a median value of 4500 μg/l. In contrast, rivers monitored by GEMS outside Europe have a much lower median value of 250 μg/l. The median phosphate level in rivers monitored by GEMS is 2.5 times the average for unpolluted rivers (10 μg/l). As regards metals and toxic substances, regulatory measures have led to a marked decrease of lead in most OECD rivers since 1970 (6). Trends in other metals and toxic substances are less encouraging, despite efforts to reduce discharges. Such substances are often persistent, accumulate in bottom sediments and can be released over long periods of time once initially deposited. Organochlorine pesticides measured in some rivers from developing countries (such as Tanzania, Colombia and Malaysia) are higher than those recorded in European rivers.

Impacts of mismanagement and pollution

Water use has not been efficient in many countries. Over-exploitation of groundwater (mostly a non-renewable source) has led to the depletion of resources in some areas, and to increased encroachment of saline waters into aquifers along coastal zones in some countries, particularly in North Africa and the Persian Gulf. There are fears that the rapid expansion of agriculture in desert areas may lead to over-exploitation of groundwater used for irrigation (7). Excessive irrigation has also led to waterlogging and salinization, which has accelerated land degradation (Chapters 6 and 11).

The lack of maintenance of water delivery systems and over-use of water for domestic, commercial and industrial purposes, especially in the developing countries, have caused a host of socio-environmental and economic problems. Pools of water around faulty standpipes in rural areas and marginal settlements have become breeding grounds for disease vectors. Water seepage has affected the interior and exterior of houses, historical buildings and monuments, and in some areas has caused occasional overflow of drainage and sewage systems. Water losses, which can amount to more than 70 per cent of water delivered, put increasing and costly pressures on water works which have to meet the increasing demand for pure water.

The quality of fresh water depends not only on the amount of waste generated but also on the decontamination measures that have been put into effect. Although organic waste is biodegradable, it nonetheless presents a significant problem, especially in developing countries. Human excreta contain such pathogenic microorganisms as the waterborne agents of cholera, typhoid fever and dysentery, and contaminated water has caused epidemics of these diseases in several developing countries (Chapter 18).

Industrial waste may include heavy metals and many other toxic and persistent chemicals not readily degraded under natural conditions or in conventional sewage-treatment plants. Unless these wastes are adequately treated at source, or prevented from discharge into water courses, freshwater quality can be seriously impaired. The high content of nutrients in rivers and lakes has created eutrophication. Apart from ecological and aesthetic damage, eutrophication brings increasing difficulties and costs for water treatment works. Acidification of freshwater lakes has affected aquatic life (Chapter 1). In most newly industrializing countries, both organic and industrial river pollution are on the increase and decontamination efforts are often neglected. In these countries, industrialization has had higher priority than the reduction of pollution. As a result, in some regions (East Asia, for example), the degradation of water resources is now considered the gravest environmental problem (3). In many of these countries, aquatic life (especially fisheries) has been affected and the deterioration of water quality is a growing threat to aquaculture, a major source of fish for the population (Chapter 11).

... in some regions (East Asia, for example), the degradation of water resources is now considered the gravest environmental problem.

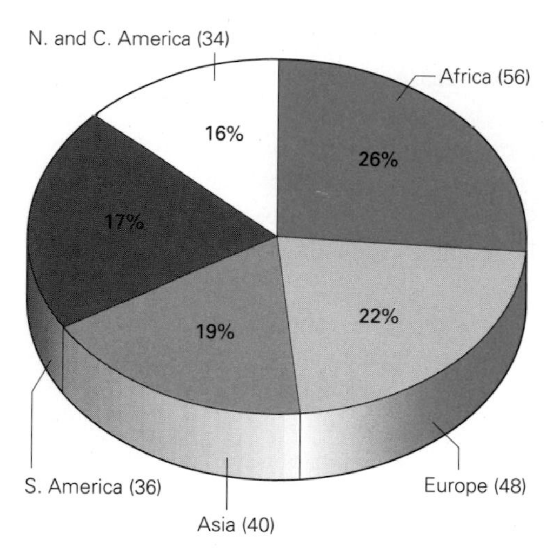

Figure 5.3
Numbers of international river basins

based on data from (17)

Shared water resources

Many freshwater resources are shared by two or more states. At least 214 river basins are multinational: 155 of these are shared between two countries; 36 among three countries; and the remaining 23 among 4 to 12 countries (Figure 5.3). About 50 countries have 75 per cent or more of their total area falling within international river basins, and an estimated 35-40 per cent of the world population lives in these basins (8).

The joint use of international watercourses has always depended on cooperation among the riparian states. International treaties and organizations were created to regulate the use of some shared water bodies. Historically, these treaties dealt with issues such as allocation of water shares, regulation of navigation and fishing, and construction of public works such as barrages. Only recently—especially since the early 1970s—have some of these treaties been revised to reflect growing concerns about pollution of shared water resources. For example, in the United States the Great Lakes Water Quality Agreements of 1972 and 1978 focused respectively on pollution from traditional sources, such as municipal sewers, that were causing severe eutrophication of the lower Great Lakes, and on toxic pollutants (2). In Europe, a joint programme for the rehabilitation of the Rhine's water and the management of the Rhine groundwater aquifer has been undertaken by the riparian countries since 1980. The Sandoz-Basel accident that occurred in November 1986 (Chapter 9) has prompted the Economic Commission for Europe to begin work on formulating regional conventions on the transboundary impacts of industrial accidents and on the protection and use of transboundary watercourses and lakes.

The level of the Aral Sea, shared by three republics in the Soviet Union, is falling because excessive irrigation withdrawals are reducing inflow from the catchment area. The Aral Sea level has dropped by 3 m since 1960 and, if this trend continues, it will drop another 9–13 m by the year 2000. Reduced inflow, with enhanced salinity from irrigation returns, has already increased the salinity of the Aral Sea three-fold to 1 g/litre, and by the year 2000 this is expected to rise to 3.5 g/litre. The proposed transfer of water from the Siberian rivers to the region would minimize the problems in the Aral Sea basin (5).

This large-scale transfer of water will ameliorate a deteriorating environmental situation and will further agricultural and economic development in the area. However, not all large systems of water transfer have net beneficial affects, and the economic, social and environmental consequences of such transfer schemes should be carefully evaluated (9).

Responses

The oldest approach to water management is the construction of dams and reservoirs to control floods and to store water for use as the need arises. Hundreds of thousands of dams and reservoirs have been built but these include only a few hundred large multipurpose dams for water management and electricity generation.

Some 36 240 dams higher than 15 metres were constructed between 1950 and 1986, 79 per cent of which are between 15 and 30 m high, 16 per cent between 30 and 60 m, 3.4 per cent between 60 and 100 m, 0.9 per cent between 100 and 150 m, and 0.28 per cent higher than 150 m (10, 11). About half were constructed in China alone. Although these dams have provided several benefits, they have not been without environmental costs, and the past two decades witnessed wide-ranging discussions about the costs and benefits of large dams, such as the Aswan High Dam and others (12, 13, 14). One important fact remains: the amount of stored water in man-made reservoirs in the world has been estimated at 3500 km^3, almost the total worldwide annual water withdrawal.

In addition to these schemes, several countries have taken measures to improve the efficiency of water use. Technical as well as regulatory measures (including pricing mechanisms, incentives and disincentives) have been introduced with varying success. In spite of these efforts, water use is still inefficient, particularly in developing countries, many of which provide water (especially for irrigation) either free or heavily subsidized. The past two decades also witnessed increasing efforts to recycle water for use in industry and agriculture (Chapters 11 and 12).

Global concern about the availability and quality of water was highlighted at the United Nations Water Conference held in Mar del Plata, Argentina, in

One important fact remains: the amount of stored water in man-made reservoirs in the world has been estimated at 3500 km^3, almost the total worldwide annual water withdrawal.

1977. The recommendations of the conference covered eight major areas: assessment of water resources; water use and efficiency; environment, health and pollution control; policy, planning and management; natural hazards; public information, education, training and research; regional cooperation; and international cooperation. The implementation of the Mar del Plata Action Plan has been rather slow, but recently the United Nations has embarked on the formulation of a global strategy to implement the Plan in the 1990s.

The Mar del Plata Action Plan and the United Nations Conference on Human Settlements, held in 1976 in Vancouver, Canada, set the stage for the launching of the International Drinking Water Supply and Sanitation Decade (IDWSSD 1981–90) by the General Assembly of the United Nations in 1980, at the recommendation of the World Health Organization. The main objective of the Decade was to bring about a substantial improvement in the standards and levels of services in drinking water supply and sanitation by the year 1990.

In 1970, 33 per cent of the population in urban areas of the developing countries did not have access to safe clean water and 29 per cent did not have access to sanitation services; in rural areas, 86 per cent did not have access to clean water and 89 per cent did not

IDWSSD and beyond

- About 1348 million people were provided with a safe drinking water supply in developing countries (368 million in urban areas and 980 million in rural areas).
- About 748 million people were provided with suitable sanitation services (314 million in urban areas and 434 million in rural areas).
- Overall, the number of people without safe water decreased from 1825 to 1232 million, while the number of people without suitable sanitation remained virtually the same.

The rate of progress achieved during the IDWSSD would be insufficient to reach the ultimate objective of services for all by the end of the century.

If programme implementation were to continue at the current rate, those in urban and rural areas unserved with safe water by the year 2000 would decrease to around 767 million due to significant increases in coverage in rural areas. In percentage terms this would constitute a decrease from 31 per cent of the total population of the developing countries in 1990 to 16 percent by 2000. Those unserved with sanitation would rise to around 1880 million, although the percentage of the population without services would decrease from 43 to 38 per cent due to a small decrease in the number of people in rural areas without coverage. The health and environmental consequence associated with these numbers of people without services would preclude the achievement of living conditions compatible with sustainable development.

Sources: (15, 16)

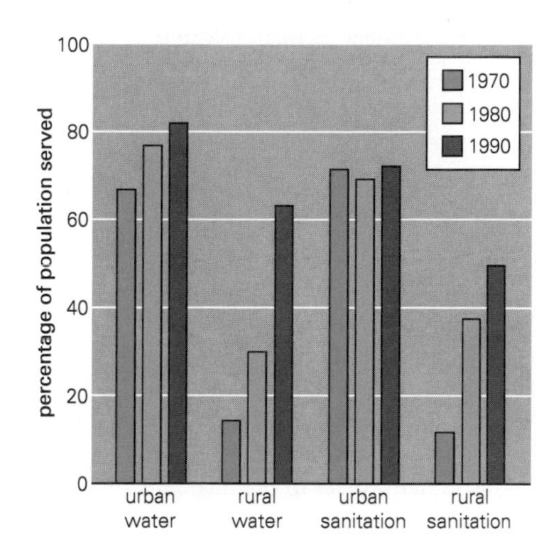

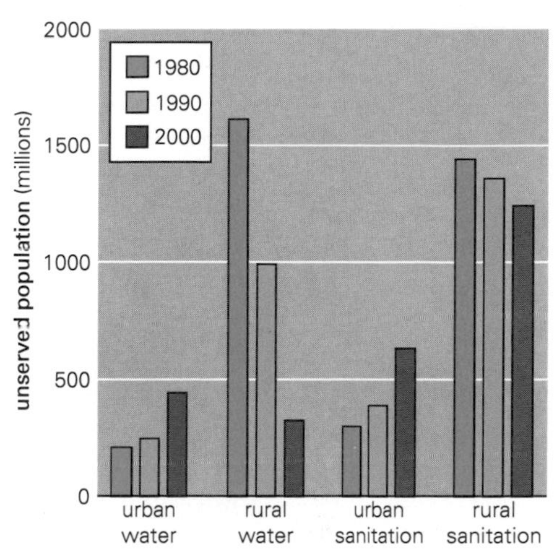

Figures 5.4 and 5.5 Water supply and sanitation coverage in developing countries (above) and numbers of people without adequate water supplies and sanitation (above right)

based on data from (15, 16)

have access to sanitation services. By the end of IDWSSD, only 18 per cent of people in urban areas were without access to safe clean water. Access to sanitation services did not improve—barely one per cent more were supplied with sanitation services during the decade. In rural areas, however, improvements were dramatic: the percentages fell to 37 without clean water and 51 without access to sanitation services (Figures 5.4 and 5.5, and box opposite).

The slow progress towards achieving the goals of the IDWSSD, particularly in urban areas, has been attributed to several factors, including population growth, rural-urban migration, the unfavourable world economic situation and the debt burden of developing countries, which has been a major obstacle to investment in infrastructure projects. However, enough knowledge and experience has been gained to reach the goal of IDWSSD by the end of the century, provided that adequate investment is made available coupled with the provision of appropriate low-cost technologies and wider public participation (box opposite).

New and more comprehensive approaches to water management are needed to enhance socio-economic and environmental development, especially in international basins (whether rivers, lakes or groundwater aquifers).

... enough knowledge and experience has been gained to reach the goal of providing fresh water and adequate sanitation to all by the end of the century.

The Environmentally Sound Management of Inland Waters (EMINWA) programme launched by UNEP in 1986 is one of these comprehensive approaches. The programme is designed to assist governments to integrate environmental considerations into the management and development of inland water resources, with a view to reconciling conflicting interests and ensuring the regional development of water resources in harmony with the water-related environment throughout entire water systems.

Within the framework of EMINWA, the Zambezi Action Plan (ZACPLAN) for the environmentally sound management of the common Zambezi river system was adopted in 1987. Eight countries—Angola, Botswana, Malawi, Mozambique, Namibia, the United Republic of Tanzania, Zambia and Zimbabwe—are participating in ZACPLAN. Another project in the final stages of development is a master plan for the development and environmentally sound management of the natural resources of the conventional Lake Chad basin area, which covers parts of Cameroon, the Central African Republic, Chad, Niger and Nigeria. Other activities under way include the management of the Nile river basin.

Another project in the final stages of development is a master plan for the development and environmentally sound management of the natural resources of the conventional Lake Chad basin area, which covers parts of Cameroon, the Central African Republic, Chad, Niger and Nigeria.

Chapter 6

Land degradation and desertification

Of the total land area in the world (about 13 382 million ha, 13 069 million of which are ice-free), only 11 per cent (about 1475 million ha) is currently under cultivation, while 24 per cent is permanent pasture, 31 per cent comprises forests and woodlands and 34 per cent is classified as 'other land', which includes unused but potentially productive land, built-on areas, wasteland, parks, or other land not specified in the previous types (1). The world's potentially cultivable land has been estimated at about 3200 million ha, more than twice the area currently used as cropland. About 70 per cent of potentially cultivable land in the developed countries, and 36 per cent in the developing countries, is currently cultivated (Chapter 11).

Data from FAO (1) show that in the 15 years from 1973 to 1988 the total area of arable and permanent cropland in the world increased from 1418 to 1475 million ha (4 per cent), that of permanent pastures decreased slightly from 3223 to 3212 million ha (–0.3 per cent), that of forests and woodlands decreased from 4190 to 4049 million ha (-3.5 per cent), and that of 'other land' increased from 4235 to 4333 million ha (+2.3 per cent).

Human activities have radically reshaped the world's natural land cover. The often indiscriminate destruction of forests and woodlands (Chapter 7), the overgrazing of vegetation by increasing numbers of livestock, and the improper management of agricultural land have all resulted in extensive degradation.

The productivity of farmland overwhelmingly depends on the capacity of the soil to respond to management. The soil is not an inert mass, but a delicately balanced assemblage of mineral particles, organic matter and living organisms in dynamic equilibrium. Soils are formed over long periods of time, generally from a few thousand to millions of years (2). Excessive human pressure or misguided human activity can destroy soils in a few years or decades, and the destruction is often irreversible.

Of all human activities, agricultural production has had the greatest impact on soil degradation. Traditionally, farming practices were sustainable and preserved the soils on which they were based. In recent decades, however, human management of agro-ecosystems has been steadily intensified, through irrigation and drainage, heavy inputs of energy and chemicals, and improved crop varieties increasingly grown as monocultures. Although it has raised productivity, this process has made agro-ecosystems increasingly artificial, unstable and prone to rapid degradation (Chapter 11).

Pressure to expand the area under farming has resulted in more and more utilization of marginal land, often with detrimental consequences. Overgrazing and overcultivation on steep hillsides has led to serious soil erosion. Slash-and-burn agriculture has accelerated deforestation, which in turn has led to increased soil erosion and floods (Chapter 7). Agricultural land has been used to provide for housing, commercial and industrial development, and transport. In some countries, coastal areas and wetlands are particularly vulnerable to these human activities (Chapter 4).

Soil degradation is a complex process involving one or more of several agents: erosion as a result of physical removal by water and wind, and chemical, physical and biological changes (3). Although erosion is a natural process, its intensity has been greatly increased by human activities. The average rate of soil erosion is estimated to be about 0.5–2.0 tonnes per hectare per year (4), depending on soil type, slope and type of erosion. In the United States, 44 per cent of cropland is affected by erosion (5). In El Salvador, 77 per cent of the land area is suffering from accelerated erosion (6). In the eastern hills of Nepal, 38 per cent of the land area consists of fields which have had to be abandoned because the topsoil has washed away. In India, about 150 of 328 million ha of farmland are affected by erosion (7). Worldwide, it has been estimated that about 25 400 million tonnes of material are removed each year from topsoil by excessive erosion (5). A decline in soil fertility or even a total loss of land to agriculture, caused by increase in salinity or alkalinity, is a common problem in many parts of the world (Chapter 11).

A recent global assessment of land degradation (GLASOD) carried out by the International Soil Reference and Information Centre (ISRIC) at Wageningen, The Netherlands (8), estimates that 15 per cent of the Earth's land area has been degraded by human activities: 55.7 per cent of this land has been degraded by water erosion, 28 per cent by wind erosion, 12.1 per cent by chemical means (loss of nutrients, salinization,pollution and acidification), and 4.2 per cent by physical means (compaction, water-logging and subsidence). The main causes are overgrazing, which accounts for 34.5 per cent of the degraded area; deforestation, 29.5 per cent; agricultural activities, 28.1 per cent; over-exploitation, 7 per cent; and bioindustrial activities (waste accumulation, excessive manuring, use of agrochemicals, etc), 1.2 per cent (Figure 6.1). GLASOD has classified the degree

In the eastern hills of Nepal, 38 per cent of the land area consists of fields which have had to be abandoned because the topsoil has washed away. In India, about 150 of 328 million ha of farmland are affected by erosion.

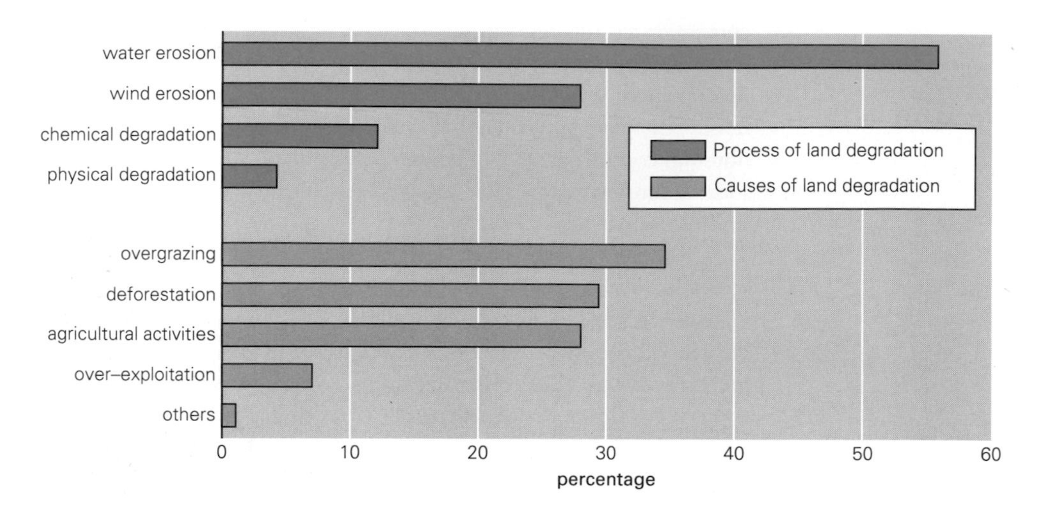

Figure 6.1 Processes and causes of land degradation

based on data from (8)

of land degradation into four categories: light, moderate, strong and extreme. According to FAO (1), the total area of agricultural land (arable land, permanent pasture and grazing land) was about 4687 million ha in 1988. The GLASOD figures indicate that around 1230 million ha of these (26 per cent) have been degraded as a result of mismanagement.

Drylands (arid, semi-arid and dry sub-humid areas) cover 6150 million ha, or about 47 per cent of the total land area in the world. They comprise 62 per cent of all irrigated land, 36 per cent of the rainfed cropland, and 68 per cent of all rangeland. Desertification, defined as land degradation in drylands resulting mainly from adverse human impact, is common in many areas. The recent assessment by UNEP of the global status of desertification (9) shows that 30 per cent of irrigated areas within the drylands, 47 per cent of rainfed cropland, and 73 per cent of rangeland are at least moderately affected (Figures 6.2, 6.3, 6.4 and 6.5). About 43 million ha of irrigated land in the world's drylands are affected by some form of degradation, mainly water-logging, salinization and alkalinization. It has been estimated that a total of 1.5 million ha of irrigated land is lost every year worldwide, of which 1.0–1.3 million ha are in drylands. Nearly 216 million ha of rainfed croplands in the world's drylands are affected by water and wind erosion, depletion of nutrients and physical deterioration. About 7-8 million ha of rainfed croplands are currently lost every year, of which 3.5–4.0 million ha are in drylands. About 3333 million ha of rangeland in drylands are affected mainly by degradation of vegetation; erosion also affects some 757 million ha of

Figures 6.2, 6.3 and 6.4 Desertification in irrigated areas, rainfed croplands and rangelands within drylands

based on data from (9)

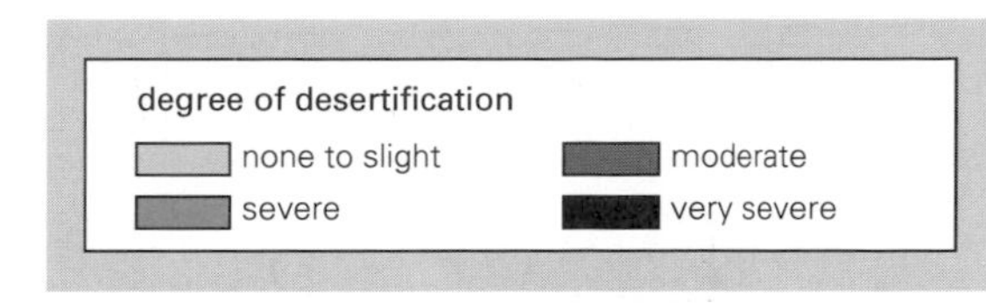

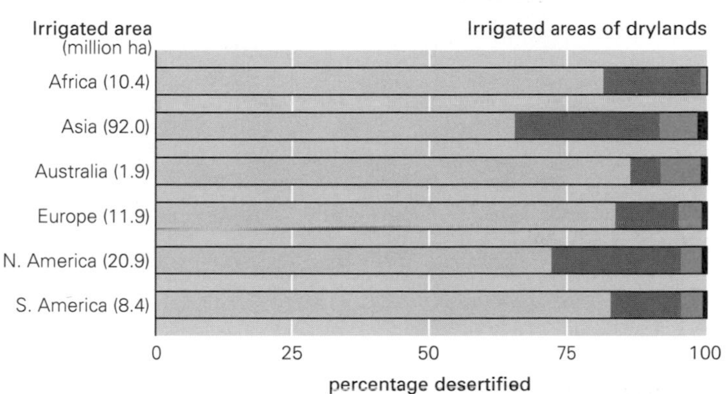

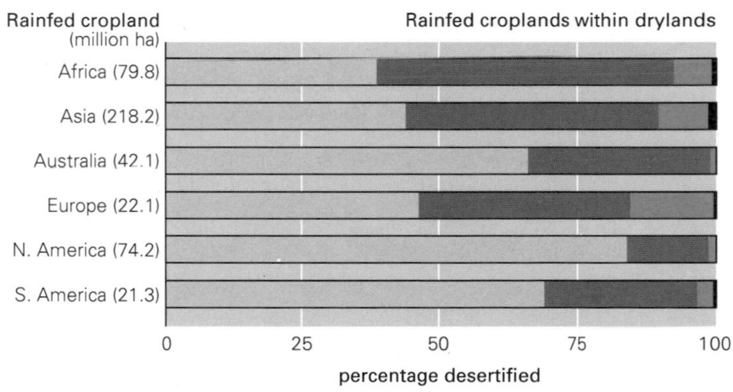

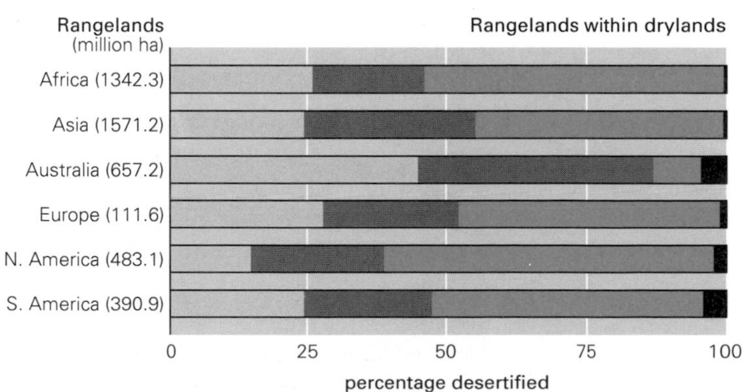

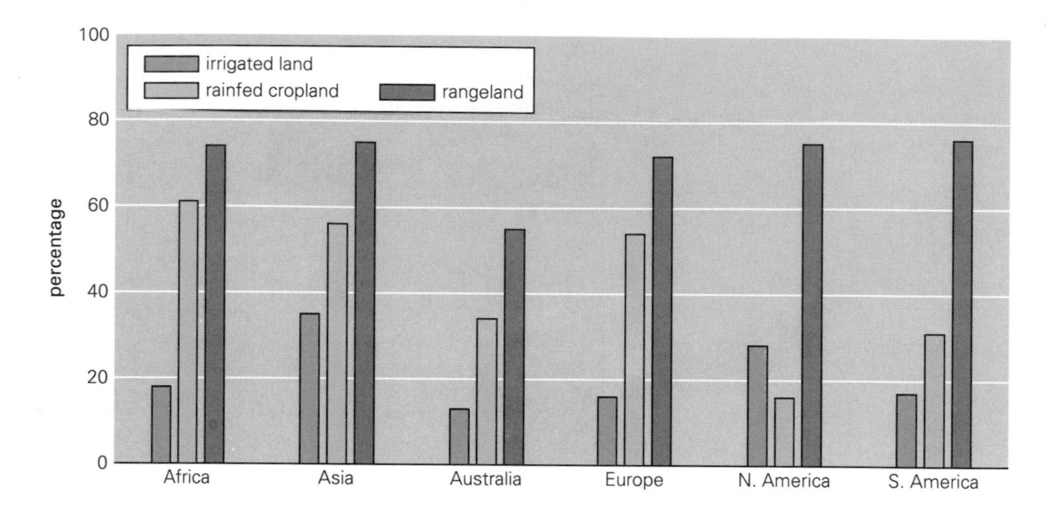

Figure 6.5
Percentage of drylands affected by desertification

based on data from (9)

this area. All in all, some 70 per cent of all agriculturally used drylands is affected by desertification/land degradation. The worst affected is North America, Africa, South America and Asia (9).

Impacts of land degradation and desertification

While people are the main agents of land degradation and desertification,they are also their victim. Throughout the developing world, land degradation has been the main factor in the migration of subsistence farmers (looking for 'better' opportunities) to the slums and shanty towns of major cities, producing desperate populations vulnerable to disease, natural disasters, the temptations of crime and civil strife. This exodus from rural to urban areas has exacerbated the already dire urban problems of many developing countries. It has also delayed efforts to rehabilitate and develop rural areas because it leads to a lack of manpower and neglect of the land.

The effects of land degradation and desertification are compounded by recurrent droughts. The mass exodus that has been taking place in Africa since the late 1970s is a vivid illustration of the plight of people facing such intolerable environmental conditions. At the peak of the crisis, in 1984 and 1985, an estimated 30–35 million people in 21 African countries were seriously affected, of whom about 10 million were displaced and became known as 'environmental refugees' (10). Death, disease, chronic malnutrition and disability are haunting these refugees because of their intolerable living conditions.

Land degradation and desertification diminish the ability of affected countries to produce food. They create food deficits in the affected regions, which have an impact on world food reserves and food trade. Since desertification entails the destruction of vegetation and diminution of many plants and animal populations, it leads to a loss of biodiversity in arid and semi-arid areas (Chapter 8), which itself limits opportunities for food production.

Responses

Preventing land degradation and desertification is certainly much more efficient and economical than rehabilitating degraded land. The latter becomes more difficult and costly as degradation advances. Many countries are undertaking costly operations. During 1976–80, more than 740 000 ha of land in Bulgaria were protected from erosion, and more than 1.4 million ha were treated to reduce soil pollution (11). In Hungary, soil erosion has caused land degradation over an area of about 2.3 million ha and efforts are under way to ameliorate the situation. Extensive drainage networks have been constructed in several countries to reduce water-logging and salinization. In Pakistan, 32 salinization control and reclamation projects were completed in the period 1960–85. As a result, the extent of salinization has declined from 40 to 28 per cent. On average, about 81 000 ha of affected land are being brought back into full production every year (12).

Efforts to rehabilitate degraded rangeland are under way in many countries. In Syria, range cooperatives have been established and regulations for the use of certain rangelands have been formulated. In Jordan priority is given to the establishment of permanent settlements for nomadic herdsmen. Several technologies are being introduced to raise the carrying capacity of rangeland. For example, it has been demonstrated in Saudi Arabia, Kuwait and Pakistan that salt-tolerant grasses grow well under irrigation with brackish water. Annual medic pasture in rotation with cereals has been utilized with varying degrees of success in Libya, Syria, Iraq and Jordan. Tropical pasture species have been introduced in Oman and Sudan where environmental conditions allow (13).

Reforestation and afforestation have been undertaken in several countries

> ***In Pakistan, 32 salinization control and reclamation projects were completed in the period 1960–85. As a result, about 81 000 ha of affected land are being brought back into full production every year.***

to stabilize soils, prevent the encroachment of sand dunes on agricultural land and halt desertification. Marked progress has been achieved in countries such as China and the Republic of Korea (Chapter 7). Agroforestry is also practised in some countries. Farmers plant trees as windbreaks or as shade trees on pastures and fields. Trees also provide fuelwood, poles, fruit, edible seeds and fodder. In North Africa, a greenbelt has been planned as part of the effort to halt desertification. In spite of these efforts, the implementation of the Plan of Action to Combat Desertification which was adopted by the United Nations Conference on Desertification in 1977 has been very slow due to several factors, the most important of which are institutional, administrative, technical and financial. A detailed evaluation of the progress achieved in the implementation of the Plan of Action to Combat Desertification is published by UNEP within another comprehensive document on desertification presented by UNEP to the UNCED.

Reforestation and afforestation have been undertaken in several countries to stabilize soils, prevent the encroachment of sand dunes on agricultural land and halt desertification. Marked progress has been achieved in countries such as China and the Republic of Korea.

Chapter 7

Deforestation and degradation of forests

Forest cover is of great ecological importance. It protects and stabilizes soils and local climates, improves the soil's ability to hold water, and increases the efficiency with which nutrients are cycled between the soil and vegetation. Forests also provide a habitat for people and numerous plant and animal species. Virgin forests, especially those in tropical regions, are an irreplaceable repository of the genetic heritage of the world's flora and fauna (see Chapter 8). Forests also provide timber and firewood, as well as medicinal and other plants of use to humankind. The role of forests as carbon sinks to reduce the effects of carbon dioxide in the atmosphere, and thereby help to contain global warming, is well-established (Chapter 3).

Forests now cover 3625 million ha, or 27.7 per cent of the world's ice-free land area (1). Of the forested area, 25.4 per cent is covered by boreal forests; 21.2 per cent by temperate forests; and 53.4 per cent by tropical forests (Figure 7.1). In addition, there are some 650 million ha covered by 'other wooded vegetation', which includes scrub, thickets, shrub vegetation, and forest fallow.

According to OECD data, the total stock of wood contained in forests can be estimated at about 315 billion m^3 (2). This is the resource base and will eventually be depleted if more than the annual increment from it, estimated at about 6 billion m^3, is consumed. In 1988, the total world consumption of roundwood was 2972 million m^3, of which industrial roundwood accounted for 1535 million m^3—or 51.6 per cent—and fuelwood for 1437 million m^3—or 48.4 per cent (3). This figure is an underestimate, since many countries keep no accurate records of self-collected or self-produced wood. Although it seems that the annual increment of the forest resource base can meet world demand for wood, forest resources are

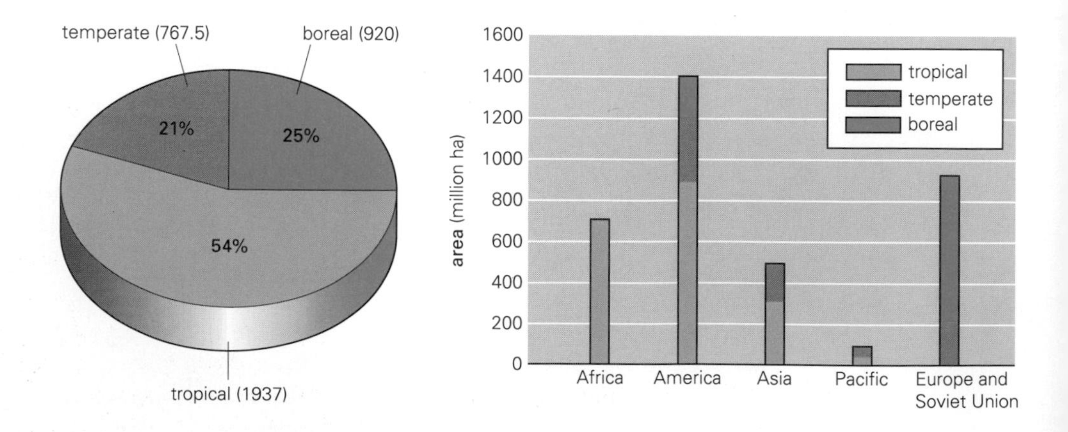

Figure 7.1 Distribution of forests by type and region (million ha)

based on data from (1)

unevenly distributed and not all the increment is being removed. Much of the latter is in the inaccessible northern forests of Alaska, Canada and the Soviet Union. This places increasing pressures on, and over-exploitation of, forest resources in other regions, such as South-East Asia and Latin America. A recent forecast (2) indicates that the supply of industrial roundwood will not meet world demand in the year 2010.

Wood is the primary source of energy for domestic heating and cooking for well over 2000 million people. Fuelwood and charcoal supplied 17 per cent of total energy consumption in developing countries in 1990 (1), but their importance is even greater in rural areas where they are the main source of energy for most households and rural industries. In some countries wood is the main source of energy, providing more than 80 per cent of energy use in such nations as the Sudan, Nigeria, Niger, Ethiopia, Mali, Nepal, Tanzania and Burkina Faso (4). Current estimates indicate than on average every person in the developing countries consumes about 0.45 m^3 of wood as fuelwood or charcoal per year, although values as high as 2.5 m^3 may be encountered in rural areas. Because of the increasing demand for fuelwood and rapid depletion of resources, in 1980 about 100 million people in developing countries could not get sufficient fuelwood to meet their minimum energy needs, and close to 1300 million consumed fuelwood resources faster than they were being replenished. Without remedial action, it is estimated that 2400 million people will be either unable to obtain their minimum energy requirements or will be forced to consume wood faster than it is being grown by the year 2000 (1). The world fuelwood deficit could reach 960 million m^3 a year by 2000.

The degradation of forests is caused by a number of natural and anthropogenic factors. Natural hazards such as droughts, frosts and storms, and the spread of pests and diseases, degrade the quality of forests in some regions. Forest fires have caused serious damage to forests in France, Greece, Spain, the United States and other countries (Chapter 9). Air pollution (especially acidic deposition and oxidants) can affect forests directly by acting on the foliage or indirectly by changing the properties of the soil supporting forest growth (Chapter 1). In 1988, it was estimated that in

... 2400 million people will be either unable to obtain their minimum energy requirements or will be forced to consume wood faster than it is being grown by the year 2000. The world fuelwood deficit could reach 960 million m^3 a year by 2000.

Europe's temperate forests between 0.6 per cent (Portugal) and 5.4 per cent (Czechoslovakia) of all trees were severely defoliated or dead. In addition, between 0.7 per cent (Portugal) and 22 per cent (Czechoslovakia) of the trees were moderately defoliated. Between 10 and 20 per cent of the trees of all species are moderately to severely affected in 13 European countries, and more than 20 per cent are so affected in three countries (5). Of a total forest area of 141 million ha in Europe, some 50 million ha (or 35 per cent) are estimated to have been damaged to some extent (6). This 'die-back' of forests has been attributed to a number of causes: acidic deposition, soil acidification, effects of atmospheric sulphur and nitrogen oxides, ozone (and possibly other photo-oxidants), climate change, pathogens and the effect of ammonium and other nitrogen compounds (7, 8, 9, 10). Air pollution has also caused forest damage in North America and East Asia. In the tropics, deforestation has been increasing due to expansion of agricultural land, ranching, and over-exploitation for fuelwood. In 1980, FAO/UNEP estimated that 11.4 million ha of closed and open forests in tropical areas were being cleared annually (11). A recent reassessment by FAO indicates that during 1981–90 the annual deforestation rate may have been as high as 16.8 million ha (1). In contrast, in temperate and boreal zones, deforestation is generally very limited and is often compensated by afforestation and reforestation.

Impacts of deforestation

Deforestation and forest degradation have many negative consequences. Tropical forests are the richest biotic environments in numbers of plant and animal species (Chapter 8). The loss of tropical forests causes the extinction of increasing numbers of these species, and forest degradation causes serious reductions in the genetic diversity of others. The loss of tropical forests already affects hundreds of millions of people through increased flooding, soil erosion and silting of waterways, drought, shortages of fuelwood and timber, and displacement of societies and cultures.

The destruction of forests undermines the basic operations of the ecosystem and may thus cause irreversible changes. The most serious of these appear to be due to the large-scale exposure of forest soils to wind and rain, leading to increased erosion and, in turn, indirectly affecting water resource development. In the past, when the Himalayas were covered with trees, Bangladesh used to suffer from overwhelming floods about once every half century. Growing populations have stripped the forests from the habitable areas on the southern slopes of the mountains. The slopes can no longer hold the rainwater, and major floods are increasing throughout the Himalayan watershed. By the 1980s, Bangladesh was suffering from a major flood about every four years. India's flood-prone area increased from 25 million ha in the late 1960s to 59 million ha in the late 1980s.

Where deforestation has eliminated plants and animals and degraded water supplies and soil fertility, families can no longer support themselves. Major deforestation can cause the displacement of whole communities. Such disruption can force people to flee and seek livelihoods elsewhere. Several million of these 'environmental refugees' have left their home countries in Central America, the Caribbean, Africa and Asia to escape poverty and environmental deterioration related to deforestation (12). In Haiti, whence more than 100 000 people have emigrated, the once abundant forests now cover less than 2 per cent of the land. In Indonesia, more than a million people have abandoned deforested and eroded areas of Java, and migrated to Borneo and other islands (14).

Deforestation has an important influence on regional and global climate. Deforestation affects regional climate by altering sensible and latent heat flux, precipitation and albedo. On a global level, deforestation has resulted in a net release of carbon dioxide and other greenhouse gases into the atmosphere. It has been estimated that tropical deforestation accounts for 26–33 per cent of the carbon dioxide released annually into the atmosphere; for 38–42 per cent of methane; and for 25–30 per cent of nitrous oxide (15). The potential for climate change (global warming) due to the increase of greenhouse gases is now well-established (Chapter 3). Although, from the theoretical point of view, such global warming could enhance the growth of tropical and temperate forests, it may have devastating effects on boreal forests (16, 17).

Responses

The logical immediate response to the growing problem of deforestation is to protect substantial areas of remaining tropical forests, to improve forest management, and to plant more trees. Worldwide, less than 5 per cent of the remaining tropical forests are protected as parks or reserves. However, Brazil has

By the 1980s, Bangladesh was suffering from a major flood about every four years. India's flood-prone area increased from 25 million ha in the late 1960s to 59 million ha in the late 1980s

Improving forest management

- Bolivia has launched a five-year ecological 'moratorium' which includes temporarily suspending logging concessions.
- The Côte d'Ivoire, which lost two-thirds of its forests in 25 years, has announced a ban on timber exports to protect its remaining 400 000 ha.
- Traditional non-destructive uses of the rainforest, such as tapping rubber and agroforestry, achieve much higher economic returns than logging, slash-and-burn agriculture and cattle ranching. Many local communities in Mexico, Brazil, Kenya, Thailand and the Philippines practise and develop such systems.

Sources (13, 14, 21, 22)

established a system of forest parks and conservation areas covering nearly 15 million ha, while Costa Rica has protected 80 per cent of its remaining wildlands through parks, wildlife refuges and reserves. Some other countries in Latin America, Africa and Asia have established successful reserves (14). Several countries have taken steps to improve forest management. Some have restricted the harvesting of timber, and others have improved harvesting technologies (see box left).

Reforestation and afforestation are under way in many countries.The FAO estimates the annual rate of successful tree planting at 1.1 million ha (1). The total area of man-made forests in the tropical countries alone was estimated to have reached 25 million ha in 1990. China is one of the few countries that has had success in reforesting major areas of its land. Between 1979 and 1983, 4 million ha were planted each year; in 1985 the area rose to 8 million ha (14). Zambia has established enough plantations to meet all its needs for industrial timber until the end of the century (13). In Cyprus, some 17 000 ha have been reforested to conserve endemic tree species (18).

Over the past two decades, some attempts have been made, with varying success, to establish 'energy farms'. Although the concept is not new (relatively large Eucalyptus energy farms, dedicated to charcoal production for steel mills, have been in operation since the early 1950s in Argentina and Brazil), it was thought that farms of fast-growing trees could meet part of the energy needs of some countries. *Ipil ipil* plantations in the Philippines were established to fuel power stations, and projections made to the year 2000 suggested that 700 000 ha of wood plantations would produce some 2000 megawatts of electricity (19). Other energy farms were established on a smaller scale in India and other countries. However, many of these energy farms have not been successful. Some of the projects were discontinued due to the drop in oil prices in the 1980s and lack of adequate funds; others were diverted and sold as timber for more profit.

Since the mid-1970s, several attempts have been made to increase the efficiency of fuelwood and charcoal utilization (which should lead to conservation of fuelwood resources). Modifications were made in wood stoves and new designs were tested (4, 20). However, the dissemination of efficient stoves has been slow and has encountered a number of economic, social and cultural problems.

Several actions have been taken to protect forests at the regional and global level. The signing of the ECE Convention on Long-range

Brazil has established a system of forest parks and conservation areas covering nearly 15 million ha, while Costa Rica has protected 80 per cent of its remaining wildlands through parks, wildlife refuges and reserves.

Transboundary Air Pollution in 1979 and the Protocols on sulphur and nitrogen oxides have led to cutbacks in emissions of sulphur and nitrogen oxides (the main agents of acidic deposition) in Europe (Chapter 1). In 1985, a Tropical Forestry Action Plan (TFAP) was launched by FAO, the World Bank, UNDP and the World Resources Institute. The Plan provides a framework for environmental management and sustainable forest development at national, regional and global levels. So far 81 countries have adopted the TFAP (1). An innovative approach introduced since 1987 is to purchase foreign debts from tropical countries in exchange for the creation of domestic forest reserves (see box below).

The International Tropical Timber Agreement which came into force in 1985 under the auspices of UNCTAD is now implemented by the International Tropical Timber Organization (ITTO), established at Yokohama in Japan in 1987. ITTO's main objectives are to improve market intelligence, assist producing countries to develop better techniques for reforestation and forest management, encourage increased timber processing in producing countries, and support research and development programmes to achieve these goals. One of the most encouraging aspects of ITTO is that producing and consuming countries are working together towards sustainable management of tropical forests. Ecological considerations have now been embedded in ITTO's objectives and activities largely due the efforts of UNEP, IUCN, WWF and many environmental NGOs.

Debt-for-nature swap

Foreign debt has been purchased at discounts of 50 to 90 per cent on the world market for several years. Generally, debt is purchased in exchange for other equity, usually funds in local currency from the debtor government.

An innovative approach introduced by non-governmental organizations is to purchase foreign debts in exchange for the creation of domestic forest reserves.

- Conservation International helped to negotiate the purchase of $650 000 worth of Bolivian debt for $100 000. In exchange, the Bolivian government committed 1.5 million ha of land and maintenance funds to expand the Rio Beni reserve.
- The World Wildlife Fund purchased $1 million of Ecuador's debt, which will be converted into funds to maintain parks and wildlife reserves.
- Costa Rica recently announced a programme to convert up to $5.4 million of its external debt, and at least eight other countries are considering similar plans.

Sources (13, 14, 21, 22)

Chapter 8

Loss of biological diversity

The Earth's genes, species and ecosystems are the product of hundreds of millions of years of evolution, and have enabled our species to prosper. But the available evidence indicates that human activities are leading to the loss of the planet's biological diversity (or biodiversity). With the projected growth in both human population and economic activity, the rate of loss of biodiversity is far more likely to increase than stabilize.

No one knows the number of species on Earth, even to the nearest order of magnitude. Estimates vary from 5 to 80 million species or more, but the figure is most probably in the range of 30 million. Only about 1.4 million of these have been even briefly described. Of these about 750 000 are insects, 41 000 are vertebrates and 250 000 are plants; the remainder consists of a complex array of invertebrates, fungi, algae and other micro-organisms (1, 2).

Like other natural resources, the distribution of living species in the world is not uniform. Species richness increases from the poles to the equator. Freshwater insects, for example, are three to six times more abundant in tropical areas than in temperate zones. Tropical regions have also the highest richness of mammal species per unit area, and vascular plant species diversity is much richer at lower latitudes (3). Between 40 and 100 species of trees may occur on one hectare of tropical rain forest in Latin America, compared to only 10 to 30 on a hectare of forest in eastern North America. About 700 species of trees have been identified in one area of about 15 hectares of rain forest in Borneo—as many as in all of North America. A region in lowland Malaysia near Kuala Lumpur has some 570 plant species greater than 2 cm in diameter per hectare (4). By comparison, the whole of Denmark possesses less than twice as many species—of all sizes—as there are in one hectare in Malaysia. Global patterns of species diversity in the marine environment resemble those on land. The number of tunicate (sea squirt) species increases from 103 in the Arctic to some 629 in the tropics. These terrestrial and marine patterns of increasing diversity in the tropics reach their peak in tropical forests and coral reefs.

Tropical forests are not, however, the only highly diverse ecosystems. Regions with a Mediterranean climate also have very rich flora, with high levels of endemism. For example, of the 23 200 species of plants estimated to occur in South Africa, Lesotho, Swaziland, Namibia and Botswana (which are temperate areas), 18 560 (80 per cent) are endemic to the region (5). This gives the area the highest species richness in the world, 1.7 times greater than that of Brazil. Some 30 per cent of California's 5046 plant species and

68 per cent of south-west Australia's 3600 plants are endemic to those regions (3).

Wetlands are among the most biologically productive ecosystems in the world, yet are often regarded as a nuisance, as wastelands, as habitats for pests and as threats to public health. In reality, wetlands help regulate water flows and provide essential breeding habitats for many species of flora and fauna. Wetlands are in retreat nearly everywhere. The United States has lost some 53 per cent of its coastal and freshwater wetlands; New Zealand has lost more than 90 per cent of its natural wetlands; and many parts of Europe have lost nearly all their natural wetlands (6). In the tropics, countries such as Chad, Cameroon, Niger, Bangladesh, India, Thailand and Vietnam have lost more than 80 per cent of their freshwater wetlands (7); the effects of this loss are felt far beyond the boundaries of any individual wetland, through disruptions of the hydrological cycle, destruction of habitats for migratory birds, and reduction of productivity of fisheries.

Loss of species

Throughout the Earth's geological history, plant and animal species were subjected to the processes of evolution. Many species became extinct during the different geological periods, which lasted millions of years. Indeed, over 99 per cent of the species that have ever existed are now extinct (8). In recent history, humans have had an increasing impact on species extinctions.

No precise estimate can be made of the number of species that have been—or are being—lost in major habitats. This is mainly due to the lack of systematic monitoring and baseline information. Many species may become extinct before they are even discovered or described. The extinction of other species may be detected years later because of inadequate monitoring. Most experts have concluded that perhaps a quarter of the Earth's total biological diversity is at serious risk of extinction during the next 20–30 years (9). Between 1990 and 2020, species extinctions caused primarily by tropical deforestation (tropical forests cover only 7 per cent of the Earth's land surface, but contain more than half of the species in the entire world biota) may

Most experts have concluded that perhaps a quarter of the Earth's total biological diversity is at serious risk of extinction during the next 20–30 years.

eliminate between 5 and 15 per cent of the world's species. This would amount to a potential loss of 15 000 to 50 000 species per year, or about 40–140 species per day (3). Data indicate that 724 species have become extinct since 1600 (2). At present, some 3956 species are endangered, 3647 are vulnerable, and 7240 are considered rare (Figure 8.1). Historically, extinction threatened mainly isolated ecosystems such as fresh water and island-dwelling species but currently 66 per cent of endangered and vulnerable vertebrates are continental.

Four main causes have been identified for the loss of species. The first is habitat loss or modification. As a general rule, reducing the size of a habitat by 90 per cent will reduce the number of species that can be supported in the long run by about 50 per cent. The second reason for the loss of species is over-exploitation. Commercial harvesting has been a threat to many marine species. Over-exploitation has been the cause of extinction of some large terrestrial animals, and well-known species such as the African elephant are under threat today. Pollution is the third reason for the growing loss of species. Pesticides have affected several species of birds and other organisms. Both air and water pollution stress ecosystems and reduce populations of sensitive species.

Figure 8.1
Numbers of extinct and threatened species

based on data from (2)

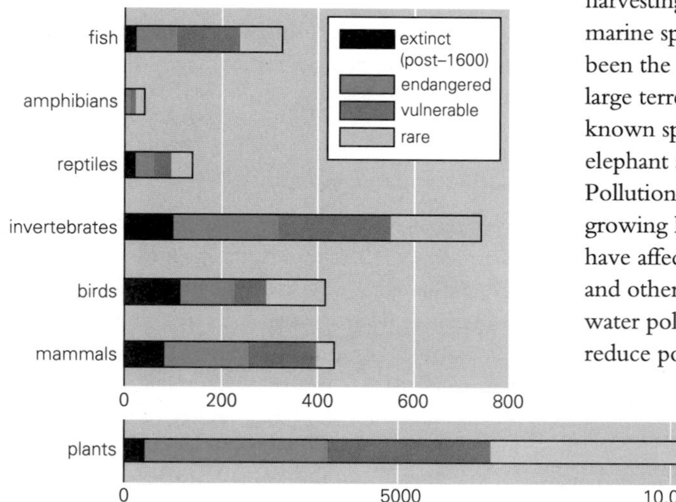

For example, air pollution and acid rain have been linked to forest diebacks in Europe and North America. Acid rain has resulted in the loss of a number of fish species in Northern European lakes (Chapter 1). The fourth reason for the loss of species is the impact of introduced exotic species which may threaten natural flora and fauna by predation, competition or altering natural habitat.

Loss of genes

A species consists of many genes; genetic diversity refers to the variation of genes within species, as expressed for example in the thousands of rice varieties in Asia. The genetic variability of many

Data indicate that 724 species have become extinct since 1600. At present, some 3956 species are endangered, 3647 are vulnerable, and 7240 are considered rare.

species is diminishing; this, in turn, diminishes their ability to adapt to pollution, climate change, disease and other forms of environmental adversity. The remaining gene pools in crops such as maize and rice amount to only a fraction of the genetic diversity that they harboured only a few decades ago. Many agriculturalists argue that the loss of genetic diversity among domestic plants and animals looms as an even greater threat to human welfare than does the loss of wild species, because that diversity is what will enable crops to adapt to future environmental change.

Impacts of loss of biodiversity

Wild species and the genetic variation within them make substantial contributions to the development of agriculture, medicine and industry. Many species constitute the foundation of community welfare in rural areas, by providing food, feed, fuel and fibre. More importantly, perhaps, many species have been fundamental to the stabilization of climate, the protection of watersheds, and the protection of soil, nurseries and breeding grounds. It is difficult to determine the economic value of the full range of good and services that biological diversity provides, but the examples in the box below are illustrative. The loss of biodiversity will restrict all these socio-economic and environmental benefits and, in the long run, will

The socio-economic benefits of biodiversity

- About 4.5 per cent of GDP in the United States (some $87 billion per year) is attributable to the harvest of wild species.
- In Asia, by the mid-1970s, genetic improvements had increased wheat production by $2 billion and rice production by $1.5 billion a year by incorporating dwarfism into both crops.
- A 'useless' wild wheat plant from Turkey was used to give disease resistance to commercial wheat varieties worth $50 million annually to the United States alone.
- One gene from a single Ethiopian barley plant now protects California's $160 million annual barely crop from yellow dwarf virus.
- An ancient wild relative of corn from Mexico can be crossed with modern corn varieties with potential savings to farmers estimated at $4.4 billion annually worldwide.
- Worldwide, medicines from wild products are worth some $40 billion a year.
- In 1960, a child suffering from leukaemia had only one chance in five of survival. Now the child has four chances in five, due to treatment with drugs containing active substances discovered in the rosy periwinkle, a tropical forest plant originating in Madagascar.

Sources (2, 11, 12)

compromise the ability of future generations to meet their needs.

Recent advances in biotechnology research and development offer new possibilities for increasing the production of food, medicines, energy, specialty chemicals and other raw materials, and for improving environmental management. This reinforces the need to maintain the richest possible pool of genes. The loss of biodiversity could cripple the genetic base required for the continued improvement and maintenance of currently utilized species and deprive us of the potential use of developments in biotechnology.

Responses

Four kinds of actions have been taken by the international community and by governments to promote the conservation and sustainable use of biological diversity : (a) measures to protect particular habitats such as National Parks, Biosphere Reserves and other protected areas; (b) measures to protect particular species or groups of species from over-exploitation; (c) measures to promote *ex situ* conservation of species in botanic gardens and in gene banks; and (d) measures to curb the contamination of the biosphere with pollutants. Several national, regional and global conventions and programmes have been formulated to implement these measures, including the Convention on Wetlands of International Importance (Ramsar, 1971), the Convention Concerning the Protection of World Cultural and Natural Heritage (Paris, 1972), the International Convention for the Regulation of Whaling (Washington, 1946), the Convention on International Trade in Endangered Species of Wild Fauna and Flora (CITES, Washington, 1973, see box below), and the Convention on

CITES

The Convention on International Trade in Endangered Species of Wild Flora and Fauna (CITES) was adopted in 1973; it entered into force on 1 July 1975. As of 31 December 1990,109 countries had become parties to the Convention.

The Treaty is designed to conserve endangered species while allowing trade in wildlife where population sizes allow. CITES bans all commercial trade related to endangered species, which it lists in its Appendix I, and limits and monitors trade related to species at risk of becoming endangered, listed in Appendix II. Appendix III allows countries to prohibit trade in nationally-protected species.

Enforcement of CITES is the responsibility of its member states, and governments are required to submit reports and trade records to the CITES Secretariat. A CITES permit is the only legal permit recognized for international transit of a wild animal, plant or product.

> ***The number of nationally protected areas increased from 1478 sites in 1970 to 6930 sites in 1990. The total area of these sites was 164 million ha in 1970 and 652 million ha in 1990. The protected areas in the world covered 4.9 per cent of the Earth's land surface area in 1990.***

the Conservation of Migratory Species of Wild Animals (Bonn, 1979). Although these have provided important means of promoting the conservation of biological diversity, none has the explicit purpose of conserving global biological diversity.

Protected areas provide a mechanism for conserving wild biodiversity, and most countries today have established at least some protected areas. The number of nationally protected areas increased from 1478 sites in 1970 to 6930 sites in 1990—about five fold (Figures 8.2 and 8.3). The total area of these sites was 164 million ha in 1970 and 652 million ha in 1990. The protected areas in the world covered 4.9 per cent of the Earth's land surface area in 1990 (10).

In response to the threats of the loss of genetic diversity, the International Board for Plant Genetic Resources (IBPGR) was established in 1974 under the umbrella of the Consultative Group on International Agricultural Resources (CGIAR). IBPGR has played a catalytic role in developing effective national and international crop genetic resource conservation efforts. Focussing largely on major crops such as wheat, rice and corn, IBPGR provided technical assistance and funding to establish national and international seedbanks and to collect a large fraction of the varieties of these crops. More recently, IBPGR has shifted its priorities to crops of regional or national importance, and has emphasized training and the building up of the human resources needed to conserve plant genetic resources.

The World Conservation Strategy (WCS) launched by IUCN, UNEP and WWF in 1980 emphasized three global objectives of living

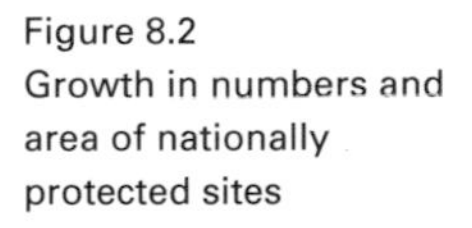

Figure 8.2
Growth in numbers and area of nationally protected sites

based on data from (10, 14)

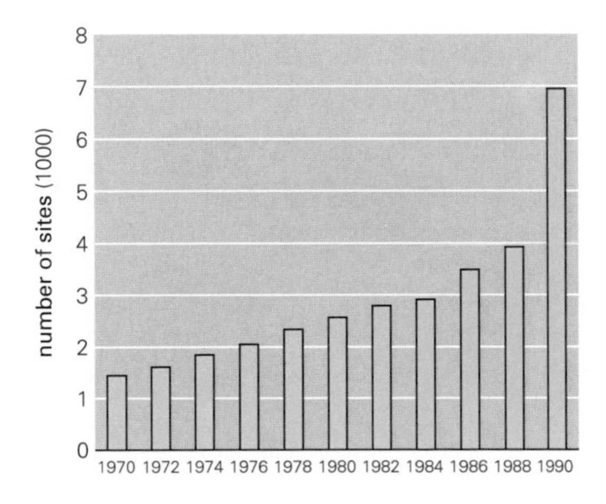

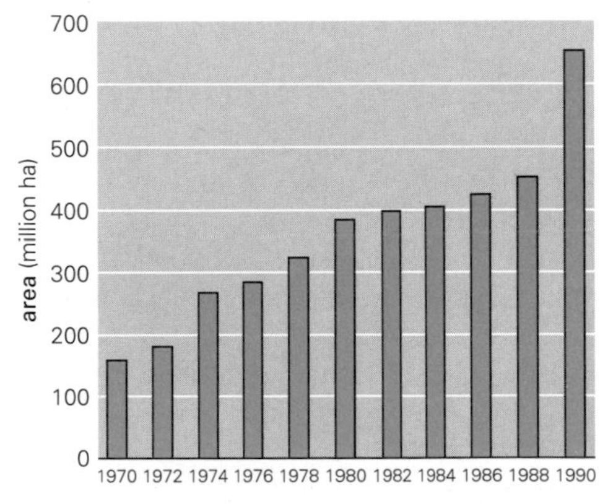

resource conservation : (a) to maintain essential ecological processes and life support systems; (b) to preserve genetic diversity; and (c) to ensure the sustainable utilization of species and ecosystems. The WCS has been used by more than 50 countries as a basis for the preparation of their national conservation strategies. The 'Caring for the Earth' strategy launched by IUCN, UNEP and WWF in October 1991 reinforces the three global objectives of 1980, and emphasizes the importance of social and economic requirements that must be met to achieve sustainable development. Caring for the Earth emphasizes that biodiversity must be conserved as a matter of principle, as a matter of survival, and as a matter of economic benefit (13).

Recognizing the growing severity of threats to biological diversity, and the increasingly international nature of the actions required to address the threats, a global strategy dealing with all aspects of biological diversity is being prepared by the World Resources Institute, IUCN and UNEP in collaboration with WWF, the World Bank, and other governmental and non-governmental institutions in both tropical and temperate nations. The strategy, to be launched in 1992, aims to: (a) establish a common perspective, foster international cooperation, and agree to priorities for international action; (b) examine the major obstacles to progress and analyse the needs for national and international policy reform; (c) specify how conservation of biological resources can be more effectively integrated with development and identify the linkages with related issues facing humanity; and (d) promote the further development of regional, national and thematic action plans for the conservation of biological diversity, and promote their implementation.

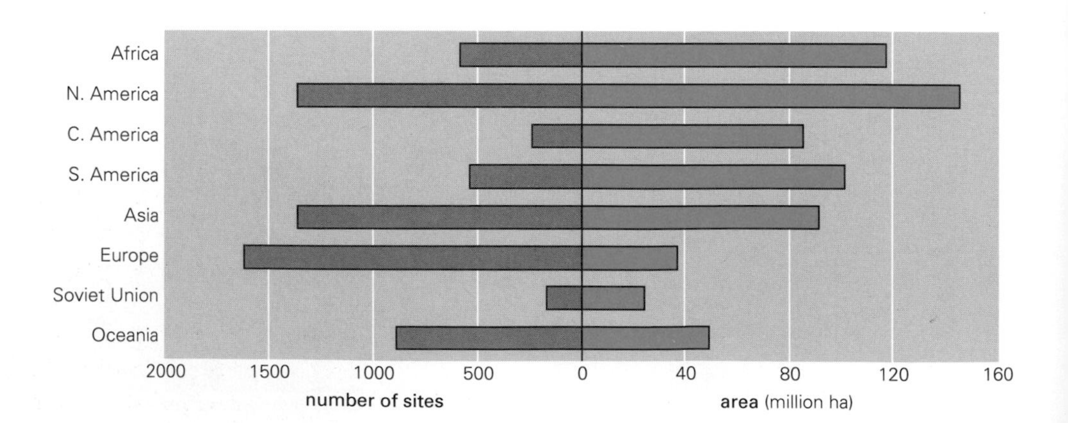

Figure 8.3
Numbers and area of nationally protected sites by region, 1990

based on data from (10, 14)

The time has come to appreciate the Earth's biological resources as assets to be conserved and managed for the benefit of all humanity. All nations have the duty to safeguard species within their territories, on behalf of everyone. But there is a need for a global effort in which developed and developing countries infuse a new spirit of cooperation for the conservation of biological diversity as a fundamental element of environmentally sound and sustainable development. UNEP, in cooperation with FAO, UNESCO and IUCN, are helping an Intergovernmental Negotiating Committee—established by UNEP's Governing Council—to elaborate an International Convention on the Conservation and Rational Use of Biological Diversity. This agreement will cover, among other things: (a) measures for conservation of the full range of biological diversity; (b) measures for sustainable utilization of biological diversity; (c) research, training, education and public awareness; (d) environmental impact assessments; (e) access to biological diversity; (f) transfer of technology—including biotechnology—for the conservation and utilization of biological diversity; (g) technical and financial cooperation with developing countries to allow them to participate fully in the conservation of biological diversity; and (h) institutional arrangements at the national and international levels.

There is a need for a global effort in which developed and developing countries infuse a new spirit of cooperation for the conservation of biological diversity as a fundamental element of environmentally sound and sustainable development.

Chapter 9

Environmental disasters

The human environment is becoming more and more hazardous. Natural disasters are becoming more frequent and catastrophic industrial accidents are on the rise.

Although natural disasters (those created by geophysical processes) are distinct from accidents (those caused by human error or technical failure), it is now recognized that human activities also enhance the occurrence and impacts of geophysical hazards. People can make land flood-prone by removing trees and other vegetation which normally absorb excess water. They can also make land more drought-prone by removing the vegetation and soil that normally absorb and store water in ways beneficial to humans. Human actions can also expose people to disasters and make them more vulnerable to their effects. In many areas in developing countries, the poor live in slums or squatter settlements unable to withstand strong wind, rain or earth tremors. Furthermore, the poor are often more exposed to the effects of industrial accidents. More and more industrial installations are built on the outskirts of cities and people move and live next to the installations because of the jobs available there or because it is cheaper to live there.

The frequency and magnitude of natural disasters have increased dramatically over the past three decades. Records of major natural disasters (1, 2) indicate that there were 16 such events in the 1960s, 29 in the 1970s and 68 in the 1980s. Although there were more natural disasters in the developed countries (63) than in the developing countries (50) during this period, they killed far fewer people (34 823 in the developed countries compared to 793 616 in the developing countries). In addition, long-lasting droughts caused the deaths of about 500 000 people between 1974 and 1984 (1), nearly all in developing countries. This illustrates the vulnerability of developing countries to the effects of natural hazards, which mostly sweep through the poorest areas. The overall economic losses due to natural disasters has also increased worldwide. Overall losses were estimated at about US$10 billion in the 1960s, US$30 billion in the 1970s and US$93 billion in the 1980s. Adjusted for inflation, the losses were an average of US$3.7 billion per year in the 1960s and US$11.4 billion per year in the 1980s (2).

Because natural disasters result primarily from geophysical interactions between the atmosphere, hydrosphere and lithosphere, changes in these interactions can alter the frequency and magnitude of disasters. Concern has been recently expressed about the possible effects of global warming (Chapter 3) on natural disasters. It is expected that the number and intensity of meteorological disasters will increase because the atmospheric heat engine will run at a higher speed if global temperatures rise. Tropical storms will become more frequent and intensive, and their path will extend increasingly toward the poles. There will be more water vapour in the atmosphere (as a result of increased evaporation rates), which will give rise to heavier rainfall, more serious floods and greater numbers of thunderstorms, hailstorms and tornadoes. In many coastal regions, the danger of storm surges will increase, especially in areas in which a rise in sea level and a greater risk of storms coincide.

Natural disasters

Although the death toll from volcanic eruptions is limited (28 666 between 1960 and 1990), few recent volcanic activities caused many deaths. An exception was the eruption of Nevada del Ruiz in Colombia on 13 November 1985 which alone caused the deaths of about 23 000 people. However, not all deaths and damage are caused by the eruption itself; the secondary effects can be almost as damaging. Volcanic ash flows, for example, can move as fast as 100 km/h down the sides of a volcano and their effects can be catastrophic if a populated area is in the path of the flow. When the loose ash becomes saturated with water from rain, it produces mud flows which are unstable and move suddenly downhill. Volcanic activity releases ash of various sizes and several gaseous products into the atmosphere. The most important of these are water vapour, hydrogen, hydrogen chloride, hydrogen sulphide, carbon monoxide, carbon dioxide, hydrogen fluoride, methane, ammonia, nitrogen and nitrogen oxides, argon and traces of mercury, arsenic and other metals. It has been estimated that volcanic activity contributes about 20 million tonnes of sulphur into the atmosphere each year in the form of SO_2, hydrogen sulphide and sulphates. This is equivalent to about 5 to 7 per cent of total global sulphur emissions into the atmosphere (3, 4).

The injection of large quantities of fine dust in the upper atmosphere from explosive volcanic eruptions could lead to climate change (5, 5, 7, 8). The US National Research Council (8) indicated that model studies predict that if one billion tonnes of ash were emitted into the stratosphere by a volcanic eruption, this would produce a worldwide drop in average

There were 16 natural disasters in the 1960s, 29 in the 1970s and 68 in the 1980s. They killed far fewer people in the developed countries (34 823) than in the developing countries (793 616).

temperature by about 10 °C for several months. However, most of the ash emitted from the eruptions of El Fuego in Guatemala in 1984 (about 7000 tonnes), Mount St Helens, Washington, in 1980 (about 100 million tonnes), and El Chichon in Mexico in 1982 (about 200 million tonnes) was dominated by large ash particles; the weight of very fine particles, which can stay in the atmosphere for longer periods, was small. The recent eruption of Mount Pinatubo in the Philippines could, however, lead to a drop in the average global temperature (see box below).

Earthquakes are the deadliest and most destructive natural disaster. Between 1960 and 1990, earthquakes killed an estimated 439 394 people and caused economic losses estimated at US$65 billion (1, 2). Although there are about one million earthquakes every year, on average only two are sufficiently strong to cause catastrophic damage. The primary effects of earthquakes are violent ground motion accompanied by fracturing which may shear or collapse large buildings, bridges, dams, tunnels, and other rigid structures. Secondary effects include short-range events, such as fires, landslides, tsunami and floods, and long-range effects, such as regional subsidence, the uplift of land masses and regional changes in groundwater hydrology.

Human activity has increased the frequency of earthquakes in three main ways. First, the earth's crust has been loaded with increasing numbers of large water reservoirs; this has created local minor earthquakes (9, 10). Second, the disposal of liquid waste in deep wells has increased fluid pressures in rocks in certain regions, producing movements along fractures. And, third, the underground testing of

Mount Pinatubo and climate

For 600 years Mount Pinatubo in the Philippines slept. On 2 April 1991 it exploded, shooting plumes of steam and ash as high as 23 km into the sky. In mid-May the volcano was emitting about 500 tonnes of sulphur dioxide daily; by 4 June this had dropped to 280 tonnes per day. After further explosions on 15 and 16 June 1991, the eruption became the largest of the century.

The ash falling on the flanks of the volcano gave rise to 'lahars' or mud flows, which swept aside everything in their way. The death toll was only around 300 because people fled their homes the first day the eruption started, but the damage to property has been very high.

Within 21 days of the first explosion, the Earth was girdled by a wide belt of fine dust and sulphuric acid aerosols that now lies between 25 °N and 20 °S. The belt covers about 40 per cent of the Earth's surface. The particulate matter in this belt reflects radiation from the sun back into space, but fails to trap heat radiating from the surface of the Earth. The net effect is cooling. According to mathematical models, the average global temperature could drop 0.5 °C for between two and four years.

Source (11)

Although there are about one million earthquakes every year, on average only two are sufficiently strong to cause catastrophic damage.

nuclear devices produces pressures within the earth which could affect the stability of parts of the earth's crust.

Earthquakes affect mostly poor people. Of the 10 deadliest earthquakes that occurred between 1960 and 1990 (see box below), 9 occurred in developing countries. Most of those who died or were injured lived in rural areas or slums. On the other hand, a number of recent earthquakes such as the 1985 Mexico City earthquake, the 1986 El Salvador earthquake and the 1988 earthquake in Soviet Armenia demonstrated that the collapse of reinforced concrete buildings is a significant problem that hinders rescue operations and may increase the death toll in urban areas.

Tropical storms (also known as cyclones, typhoons or hurricanes) are rivalled only by earthquakes as the most devastating of all natural hazards. Between 1960 and 1990, they killed 350 299 people and caused economic losses estimated at US$34 billion (2). Again, most of the devastation happened in developing countries. In Bangladesh alone two major cyclones (in 1970 and in 1985) killed about 311 000 people—89 per cent of those killed by cyclones worldwide between 1960 and 1990. A recent cyclone hit Bangladesh on 29 April 1991 killing 132 000 people. The destructive power of a tropical storm depends on three principal effects—strong winds, flooding and storm surges. The latter were responsible for most deaths in the 1970 and 1991 disasters that hit Bangladesh. Altering the environment can make people and property more vulnerable to the

The ten deadliest earthquakes, 1960–90

		number killed
27 July 1976	China	242 000
31 May 1970	Peru	67 000
7 December 1988	Soviet Union	25 000
4 February 1976	Guatemala	22 778
16 September 1978	Iran	20 000
9 February 1960	Morocco	13 100
19 September 1985	Mexico	10 000
10 April 1972	Iran	5400
23 December 1972	Nicaragua	5000
24 November 1976	Turkey	3626

Source (1, 2)

effects of tropical storms. The destruction of coral reefs, mangrove and other sea front forests, and the levelling of beach dunes all clear paths which allow storm surges to reach people and their property more quickly and forcefully.

Floods occur in many countries, developed and developing, and are nearly annual events (12, 13). Although many cause no deaths, others kill hundreds of people. Between 1960 and 1990, severe floods caused the death of about 6592 people worldwide (2). Economic losses were conservatively estimated at US$50 billion between 1970 and 1990. In OECD countries, the estimated damage from floods during the period 1975–90 was about US$9 billion (12). Although hundreds of millions of people have been affected by floods, people continue to inhabit flood plains—indeed they are occupying such areas with increasing intensity. They have altered their physical environment to suit their purposes and, in so doing, have frequently established conditions that generate more severe flooding.

Drought is the most complex and least understood of all natural disasters, affecting more people than any other disaster. The World Commission on Environment and Development (14) estimates that more than 40 million people in Africa were affected by drought during the 1980s, compared to the 24 million affected worldwide during the 1970s. Although droughts in the 1980s in Africa, China, South and South-east Asia, and South America showed how vulnerable developing countries are, recent droughts in the United States, Australia and Canada emphasize that all nations are vulnerable.

It is difficult to estimate the economic, social and environmental costs of droughts which depend, among other things, on the socioeconomic conditions of those affected and on the length of the drought. It has been estimated that the 1988 drought in the United States caused damage and losses of about US$40 billion (15).

Drought results mainly from special fluctuations of atmospheric circulation. Recently, it has been suggested that the El Niño Southern Oscillation (ENSO) has been associated with droughts in various parts of the world (El Niño is the temporary invasion of warm sea surface water into the eastern equatorial Pacific accompanied by oscillation of mean pressure differences between the western and the eastern equatorial Pacific). The ENSO of 1982–83 was the most intense for at least a century. It is claimed that it was largely responsible for droughts that occurred in 1982–83 in Indonesia, Australia, India, the United States and several African countries (16).

The United States drought in 1988 may have been linked to the El Niño of 1987 (17). Human activities may enhance the incidence of drought or increase its duration. There is speculation that the persistence of drought in western Africa is due to a combination of atmospheric circulation fluctuations plus changes induced by human activities (18). Overgrazing and deforestation—which remove vegetative cover—affect the surface albedo (radiation budget), surface roughness and moisture recycling mechanisms, thereby augmenting droughts.

Drought affects environment in a number of ways. The most pervasive effect is on soil conditions. Long periods of drought cause severe desiccation and upset the biological reactions in soils, leading to their deterioration and the enhancement of desertification (Chapter 6). Desertification is attributed in part to droughts, overgrazing and firewood cutting. Substantial evidence indicates that drought promotes outbreaks of plant-eating fungi and insects (19), and this exacerbates the already poor condition of plants.

The most dramatic effects of drought are on people. Pastoralists are often the first to feel the impact of a drought. In the Sahel, repeated droughts drove hundreds of thousands of nomadic pastoralists southwards after they had consumed the last shreds of dried-up vegetation. Many of these 'environmental refugees' moved to coastal west African nations where they took menial jobs and swelled shanty towns and slums (20). Because of long and recurrent droughts in early 1984, more than 150 million people in 24 western, eastern and southern African nations were on the brink of starvation. Ethiopia and Somalia were the most seriously affected countries in eastern Africa. Governments, trying to cope with the starving populations, established hundreds of transit and refugee camps which had to rely heavily on assistance from the international community. But many of the refugees—women, children and the elderly—did not survive their journeys. Starvation, dehydration and infectious diseases combined to accelerate the death of hundreds of thousands. Conservative estimates indicate that the death toll in Africa directly linked to droughts was about 500 000 between 1974 and 1984 (1).

The World Commission on Environment and Development estimates that more than 40 million people in Africa were affected by drought during the 1980s, compared to the 24 million affected worldwide during the 1970s.

Human-induced disasters

Forest fires—or wildland fires—are caused by natural causes (such as lightning), or by human action (negligence, accidents and arson). In densely populated regions, the

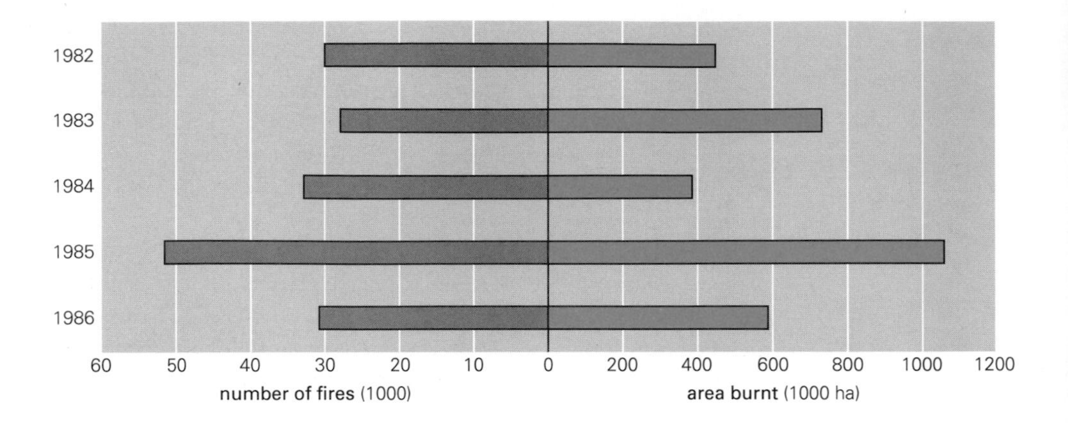

Figure 9.1
Forest fires in the Mediterranean area

based on data from (21)

latter causes are now preponderant compared with natural causes. Fire is the main cause of forest destruction in the countries of the Mediterranean basin (Figure 9.1). About 50 000 fires sweep through 700 000 to 1 000 000 hectares of Mediterranean forest each year, causing enormous economic and ecological damage as well as loss of human life (21). Most of these fires were set by people, although lightning was the cause of a fire that burnt more than 30 000 ha in Ayora-Enguera, Spain, in 1979. The economic losses due to forest fires ranged between US$17 million in Portugal to US$111 million in Spain in 1985.

Wildfires also burn millions of hectares of African savannah each year. In Asia, a single fire in Kalimantan, Indonesia, damaged more than 3.6 million ha in 1982. In North America, notwithstanding extensive, highly sophisticated prevention and control efforts, more than 2.3 million ha of forest land still burn each year. During 1988, nearly 75 000 fires burned more than 2 million ha of wildland in the United States (22).

Besides the economic losses incurred—and in some fires loss of life as well—forest fires have a number of environmental impacts. The first is on soil, where effects vary greatly, depending on the duration, extent and intensity of the fire, and on soil characteristics. Burning increases nitrogen fixation in the soil. Available phosphorus levels are increased on sandy soils and basic cations are released which may have a significant impact on the effects of acid rain by neutralizing the acidic components in precipitation (21, 23).

Forest fires emit a number of gases into the atmosphere. In 1988, forest fires in the United States contributed about 1.7 million tonnes of particulate matter, 13.6 million tonnes of carbon monoxide

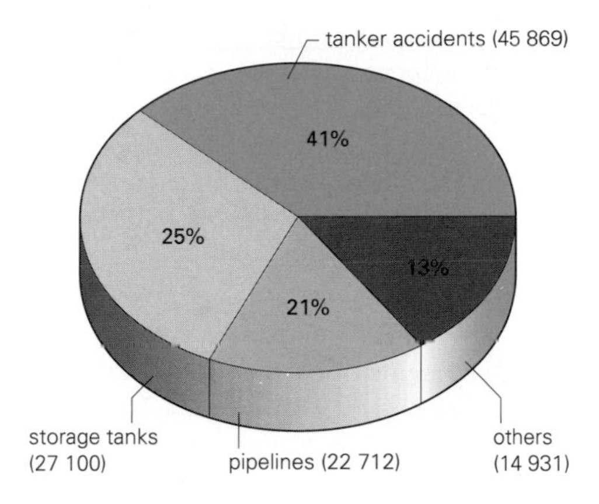

Figure 9.2
Oil spills from accidents (tonnes, 1990)

based on data from (24)

and 0.7 million tonnes of methane to the atmosphere (23). Although the direct impacts of such air pollutants may be restricted to local areas, they contribute to the global budget of such emissions.

Accidental releases of oil (and oil products) occur on land and in the sea. The latter are the ones that normally make headline news, although data show that oil spills on land are equally significant. For example, in 1989 about 239 000 tonnes of oil were accidentally spilled worldwide. Of these 185 000 tonnes (or 77 per cent) were spilled because of tanker accidents at sea. In 1990, about 111 000 tonnes of oil were spilled, of which 46 000 tonnes (or 41 per cent) were spilled from tanker accidents at sea (24). Accidental oil spills on land occur mainly from storage tanks and pipelines (Figure 9.2). Recent data show that the oil spilled from tanker accidents decreased from an average of 0.2 million tonnes per year in the early 1970s to about 0.11 million tonnes per year in the late 1980s (25). This has been partly attributed to a decrease in oil transportation by sea and partly to improvements in safety measures of tanker operations. In 1980, 1319.3 million tonnes of crude oil were transported by sea; by 1989, this figure had dropped about 20 per cent, to 1097.0 million tonnes (25).

More than 1000 tanker accidents occurred in the period 1970 to 1990. About 75 per cent of them spilled between 50 and 5000 barrels of oil each (6.9–694 tonnes); the rest spilled more than 5000 barrels each (Figure 9.3). The cumulative amount of oil spilled into the ocean between 1970 and 1990 from tanker accidents (involving spills of

Figure 9.3
Incidence of oil spills from tanker accidents

based on data from (13, 27)

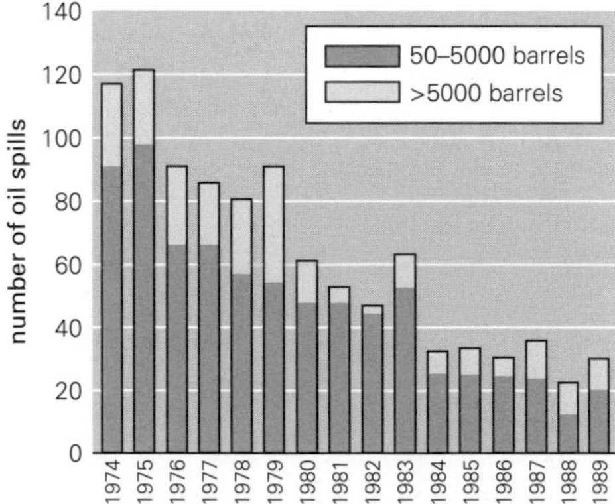

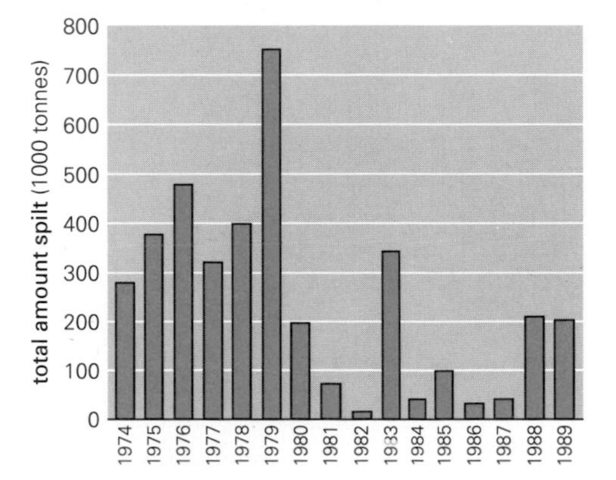

more than 25 000 tonnes each) was about 3 million tonnes (12, 25). Ten major accidents accounted for about half of this (see box below).

Another source of oil spills in the sea is from accidents at offshore rigs. *The Ekofisk* blow-out (North Sea) on 22 April 1977, caused a spill of about 15 000–21 000 tonnes. The blow-out of *Ixtoc I* in the Gulf of Mexico in 1979 spilled about 475 000 tonnes over a period of 290 days, and was the largest oil spill in the past two decades.

The extent of the damage caused by oil spills depends on several factors including the location of the spill (near shore or in the open ocean), the weather and the composition of the oil. Although there is no evidence that oil spills from tanker accidents have caused damage to the open ocean and its living resources, spilled oil can affect coastal zones, where oil may persist for several decades (26, 27). Marshes and mangroves are especially vulnerable. Wildlife is often the most conspicuous victim of oil spills. Many biological effects induced by hydrocarbons have been measured and some patterns are apparent: generally, young life stages are more sensitive than adult ones, and crustaceans are more sensitive than fish. Although studies of oil spills have shown that affected environments do recover with time, there is growing concern about the long-term effects of exposure of marine biota to low levels of hydrocarbons.

Accidental oil spills can be quite costly. The principal costs involved include containment, clean-up and environmental restoration costs, damages to fisheries and losses suffered in tourism (see panel right).

The ten largest spills from oil tankers, 1970–90

date	*tanker*	*country affected*	*oil spilled (1000 t)*
July 1979	*Atlantic Express*	Tobago	276
August 1983	*Castello Belver*	South Africa	256
March 1978	*Amoco Cadiz*	France	228
December 1972	*Sea Star*	Gulf of Oman	120
February 1980	*Irenes Sarenada*	Greece	102
May 1976	*Urquiola*	Spain	101
February 1977	*Hawaiian Patriot*	Hawaii	99
November 1979	*Independenta*	Turkey	95
January 1975	*Jacob Maesk*	Portugal	84
December 1985	*Nova*	Iran	71

Source (12, 24)

Between 1970 and 1990, about 180 severe industrial accidents occurred worldwide, leading to the release of various chemical compounds into the environment (see box overleaf). These accidents, caused mainly by fires, explosions or collision during transport, killed about 8000 people, injured more than 20 000 and led to hundreds of evacuations involving hundreds of thousands of people (12). Severe industrial accidents appear to be becoming more common. During 1974–78, there were five major accidents (accidents resulting in at least 100 deaths, 400 injured or 35 000 people evacuated). During 1984–88, the figure was 16 accidents (28). As long as strict safeguards and standards are not implemented, and as long as industrial installations are not located far from dense population centres, major accidents are likely to increase, particularly in developing countries. The massive explosion at the liquefied petroleum gas storage facility in the crowded San Juanico neighbourhood of Mexico City in November 1984 killed 452 people, injured 4248 and displaced 31 000. The blast illustrated the precarious nature of a city where many of the 17 million inhabitants live cheek by jowl with a variety of potentially dangerous installations. The Bhopal accident is another example; most of the Bhopal victims lived in squatter settlements near the plant where the accident occurred.

A number of administrative and technical steps have recently been taken to prevent such accidents and mitigate their consequences. One example is the European Economic Community's directive on the major hazards of certain industrial activities (the 'Seveso' directive). The directive obliges manufacturers within the Community to identify potential danger areas in the

The Amoco Cadiz

During the night of March 16–17, 1978, the supertanker *Amoco Cadiz* ran aground near the shore of Portsall in Brittany, France. Almost the entire cargo (228 000 tonnes) was lost in the sea within 14 days. Approximately 300 km of shore were polluted. The oil released was converted to a reddish-brown water-in-oil emulsion (mousse) by tide- and wave-induced mixing with water. Evaporation of the more volatile components is thought to have carried 20–40 per cent of the spilled oil into the atmosphere.

The oil spilled caused the death of some 4500 birds belonging to 33 species, but most were alcids and cormorants. A decrease in phytoplankton productivity was noticed for several weeks. There were minor effects on fisheries, particularly on young sole which grew at about 30 per cent of their normal rate.

The total cost of the spill has been estimated at about US$380 million ($1981): clean-up cost, US$142 million; loss to commercial fisheries, US$46 million; loss to tourism in the area, US$192 million.

Source (26,29)

The Exxon Valdez

On 24 March 1989, the supertanker *Exxon Valdez* ran aground on Bligh Reef, Prince William Sound, Alaska. Some 36 000 tonnes of oil were released into the water. Winds of more than 100 km/h on the third day after the grounding made it nearly impossible to contain the spread of oil, and within weeks about 2000 km of shoreline in south-central Alaska—a pristine environment—were affected by oil.

There were more than 10 million sea birds, 30 000 sea otters and 5000 bald eagles in the affected area. Between March and September some 36 000 birds, about 1000 sea otters and 153 eagles were killed by the spilled oil.

Although the spill disrupted the herring and salmon harvests of commercial and subsistence fisheries that had consistently supported people in the area, no marked effects from the spill were detected on herring spawning or on pink salmon. No effects were detected on inter-tidal plants; and average hydrocarbon concentrations in the water have been consistently below State of Alaska standards and were 10–100 times lower than those lethal to plants and animals living in the water.

Field counts of plants, fish and mammals from throughout the spill area have shown that wildlife species are surviving and reproducing, thus confirming that biological recovery is rapidly taking place.

The clean-up programme of the oil spill cost about US$2 billion.

Source (30, 31)

manufacturing process and to take all necessary measures to prevent major accidents as well as to limit their consequences—should they occur—for man and the environment.

In 1988, UNEP launched the Awareness and Preparedness for Emergencies at Local Level (APELL) programme to alert communities to industrial hazards and to help them to develop emergency response plans, through the dissemination of information, training, exchange of information and assistance in case of an emergency. The ILO has recently issued a code of practice to provide guidance in setting up an

Chemical accidents that made headline news, 1970–90

Seveso
On 10 July 1976 an explosion at the ICMESA chemical factory in the North Italian town of Seveso released a cloud of chemicals into the atmosphere contaminating the surrounding area.

The chemicals contained 2 kg of dioxin, a potentially toxic compound. The cause of the accident is believed to have been a 'runaway' reaction in the reactor producing sodium trichlorophenate, a main product.

There were no deaths but 200 people suffered slight injuries. The main victims were domestic animals. Contamination of the land affected some 37 000 people. Restrictions were imposed for six years over an area of 1800 ha. The worst affected area covered 110 ha. The estimated direct costs of the accident were about US$250 million.

Source (28, 32, 33, 34)

Bhopal
On the night of 2/3 December 1984, a sudden release of about 30 tonnes of methyl isocyanate (MIC) occurred at the Union Carbide pesticide plant at Bhopal, India. The accident was a result of poor safety management practices, a poor early warning system and a lack of community preparedness.

The accident led to the death of more than 2800 people living in the vicinity and caused respiratory and eye damage to more than 20 000 others. At least 200 000 people fled Bhopal during the week after the accident

Estimates of the damage vary from US$350 million to US$3 billion.

Source (28, 35, 36, 37)

Basel
On 1 November 1986 a fire broke out at a Sandoz storehouse near Basel, Switzerland. The storehouse contained about 1300 tonnes of at least 90 different chemicals. Most of these chemicals were destroyed in the fire but large quantities were introduced into the atmosphere, into the river Rhine through run-off of fire-fighting water (10 000–15 000 m^3) and into the soil and groundwater at the site In all between 3 and 30 tonnes of chemicals that entered the Rhine.

Biota in the Rhine were heavily damaged over several hundred kilometres. Most seriously affected were the benthic organisms and the eels, which were completely eradicated for a distance of about 400 km (an estimated 220 tonnes of eels were killed).

Within a few months the river Rhine had purged itself of all the chemicals released from the accident (with the possible exception of mercury and endosulfan). One year after the accident most aquatic life returned to the situation that existed before the accident. However, the groundwaters in the extensive Rhine alluvial aquifer are still polluted.

The damage caused by the Basel accident has been estimated at US$50 million.

Source (28, 38)

administrative, legal and technical system for the control of major hazard installations (39). The Basel accident (see box opposite) has made it clear that industrial accidents can have harmful transboundary impacts. This has prompted the Economic Commission for Europe to begin work on formulating a regional convention on the transboundary impacts of industrial accidents.

As of 31 December 1990, there were 423 nuclear reactors operating in 24 countries worldwide, 112 of which are in the United States (40). 'Routine' accidents—referred to as 'unusual events'—frequently occur during the operation of these reactors. These unusual events are classified by the International Atomic Energy Agency (IAEA) into events unrelated to safety (with an average frequency of 0.5 to 1 event/week/reactor), safety-related events (0.5 to 1 event/month/reactor), and events of safety significance (0.5 to 1 event/year/reactor) (41). Although the IAEA established an Incident Reporting System (IAEA-IRS) in the early 1980s, reporting of such events to IRS has been rather uneven and incomplete. The adoption in 1986 of the Convention on Early Notification of a Nuclear Accident should improve that situation. The convention entered into force on 27 October 1986 and, as of 31 December 1990, 49 countries had signed it.

During 1970–90, several unusual events led to reactor shutdowns. In the United States, for example, 16 events led to reactor shutdowns in two months—May and June 1976 (42). Forty-four events led to reactor shutdowns between August and December 1982 (43, 44) and, between May and September 1984, there were 195 such events in the United States (45). In general, these and similar shutdowns did not result in the release of radioactivity into the environment (although a few did lead to the contamination of some workers and/or restricted areas around the plants). Human error caused most of these unusual events (46).

A catastrophic reactor accident would include a complete loss of cooling, melting of the nuclear core, breaching of the reactor pressure vessel or piping, failure of the primary containment, and release of significant quantities of radioactive materials. Several studies have tried to establish the probabilities of reactor accidents of various degrees of severity. The much publicized Reactor Safety Study

In the United States, 16 events led to reactor shutdowns in two months—May and June 1976. Forty-four events led to reactor shutdowns between August and December 1982 and, between May and September 1984, there were 195 such events.

The accidents at Browns Ferry, Three Mile Island and Chernobyl

Browns Ferry

On 22 March 1975, the Browns Ferry power plant in the United States was subjected to a fire that lasted seven hours. The fire was begun by a small lighted candle being used to check for air leakage of the reactor containment building. The flame ignited some polyurethane used to seal leakage paths.

The damage inflicted by the fire resulted in the loss of the emergency core-cooling system. But alternatives were available, adequate cooling was provided throughout the event and the reactors were shut down. No one on site was seriously injured. No radioactivity above normal operating amounts was released, and there were no adverse effects on public health and safety.

The direct cost of the accident was estimated at US$10 million. The additional costs of finding replacement power raised the total cost of the accident to about US$150 million.

Source (51)

Three Mile Island

Early on the morning of 28 March 1979, the 880 MWe Three Mile Island Unit 2 (TMI-2) pressurized water reactor—which was operating at nearly full power—experienced a loss of feedwater supply that led to a turbine trip and later to a reactor trip. A series of events then caused serious damage to the reactor core. In places, core temperatures rose high high enough to melt the fuel. The accident occurred because of a combination of design, training, regulatory and mechanical failures, and human error.

From 28 March to 7 April 1979, radioactive fission products were released into the environment. The release consisted mainly of noble gases (xenon 133, xenon 135) and traces of iodine 131. Approximately eight per cent of the core inventory of xenon 133 was released. Another major release of noble gas radionuclides occurred during the controlled purge of the reactor building about 15 months after the accident. Approximately 46 per cent of the krypton 85 inventory was discharged into the atmosphere.

No one was killed as a result of the accident and there were no noticeable public health effects from the radiation released. The accident led to the evacuation of some 220 000 people from around the site, for varying periods of time. Some stress and psychological disorders were reported among the population.

The total cost of the accident was at least of US$2 billion.

Source (50, 52)

Chernobyl

The Chernobyl nuclear power station near Pripyat, 130 km north of Kiev in the Soviet Union, consisted of four units of water-cooled graphite-moderated reactors. On 26 April 1986, an explosion occurred at unit 4. The accident resulted in fuel fragmentation, and steam and hydrogen explosions; the temperature of the burning reactor rose to several thousand °C, resulting in the meltdown of the core and a release of radioactivity over a period of 10 days. The accident began while operators were testing a turbine. Safety procedures were not followed, however. Most of the neutron-absorbing control rods had been withdrawn and a chain reaction—as in a nuclear bomb—was narrowly avoided.

The accident led to a substantial atmospheric release of radionuclides. About 30 radionuclides were released with a total activity of about 2900 PBq (peta-becquerels)—about eight per cent of the total inventory of radionuclides. Among the releases were the biomedically significant nuclides strontium 90, iodine 131 and caesium 137.

The radioactive material emitted reached locations thousands of kilometres from its source. It crossed the border into Poland, southern Finland and into Norway and Sweden The extent of contamination was largely governed by whether rain washed the radioactive material from the clouds. Among other places, hot spots appeared across the Soviet republics, Scandinavia, Britain, southern Germany and Greece.

Initial concern focused on iodine 131 which was taken up by grazing cows and expressed in their milk. Leafy vegetables and fruit grown out of doors were also contaminated and had to be thrown away. But iodine 131 has a half-life of only eight days, and attention soon switched to the potential hazards of caesium 134 and 137. Caesium 137 has a half-life of more than 30 years. Caesium contaminated meat, so special measures were introduced in Scandinavia and Britain to restrict the movement of livestock.

Although the initial death toll of the accident was 31, it had reached between 250 and 350 four years after the accident. Medical data show that, during 1986–90, in the zone of strict control around Chernobyl, there was a 50 per cent increase in the mean frequency of thyroid disorders, malignancy and neoplasms (leukaemia rates increased by 50 per cent), as well as a serious increase in the numbers of miscarriages, stillborn babies and children born with genetic malformations.

There were several attempts to assess the health effects of the Chernobyl accident outside the Soviet Union. No acute effects were found, and projections of excess cancer risks for the northern hemisphere range from zero to 0.02 per cent.

The direct and indirect economic costs of the Chernobyl accident were very high—estimated at at least US$15 billion, 90 per cent of which would be in the Soviet Union

Sources (53-61)

(also known as the Rasmussen or WASH-1400 report), published in 1975 (47), estimated the probability of a meltdown in a pressurized water reactor at 1 in 20,000 per reactor per year, and that most meltdowns would not breach the main containment above the reactor. WASH-1400 estimated that the worst accident might happen once per 10 million years of reactor operation, and might cause 3300 early fatalities, about 10 times that number of early illnesses, additional genetic effects and long-term cancers, and perhaps some US$14 billion in property damage (47). The WASH-1400 estimates have been widely criticized, particularly because much uncertainty is involved in quantifying the risks of an accident with major consequences (48, 49, 50).

The occurrence in 1975 of the fire at Browns Ferry nuclear power plant (see box), as a result of a human error, fuelled the debate on the safety of nuclear installations and the validity of studies such as WASH-1400. Four years later, the Three Mile Island accident occurred (see box), after just 1500 years of worldwide reactor operation. Although not a light water reactor, the Chernobyl disaster in 1986 (see box) followed after another 1900 reactor years. If this 'historical' accident rate continues, three additional accidents would occur by the year 2000, at which point—with more than 500 reactors in operation worldwide—core-damaging accidents would occur every four years (62). But no one knows how often nuclear disasters will happen, and no one knows the extent of the damage that might occur to people and to the environment.

Although accidents at nuclear facilities have been responsible for the majority of deaths and radiation overexposure (Figure 9.4, page 101), accidents related to the use of radioisotopes in industry, research and medical facilities account for a significant number of casualties from radiation accidents. The number of such accidents has recently increased. For example, there were eight fatal accidents between 1970 and 1987 as compared to nine such accidents in the previous 25 years. The radiation accident at Goiania, Brazil, in 1987 (see box overleaf) has demonstrated that public awareness of the potential danger of radiation sources is an important factor in reducing the likelihood of radiological accidents, and in reducing the consequences of such accidents if they occur.

If this 'historical' accident rate continues, three additional accidents would occur by the year 2000, at which point—with over 500 reactors in operation worldwide—core damaging accidents would occur every four years.

Public perception of environmental hazards

People respond to the hazards they perceive. Their perception is conditioned by cultural, traditional, socio-economic and political factors. A common feature in both developed and developing countries is that public concern becomes highly stimulated when a significant hazardous environmental incident occurs. For example, public concern about the hazards of chemicals and nuclear power peaked following the accidents at Seveso (1976), Bhopal (1984), Basel (1986), Three Mile Island (1979) and Chernobyl (1986). This is natural, because the public perception of a hazard is heavily weighted by its severity and very little by its frequency.

The mass media play a major role in affecting and shaping public perception. Because news is (almost by definition) about the unusual, the media generally emphasize hazards that are relatively serious and/or relatively rare. Catastrophic events are reported much more frequently than less dramatic causes of death and damage with similar (or even greater) statistical frequencies. Events such as those mentioned above received extensive coverage by the press, radio and television, partly because they have inherent public appeal. For example, the Bhopal accident was ranked as the second biggest news

The radiological accident in Goiania, Brazil

On 13 September 1987, a shielded, strongly radioactive caesium 137 medical source was stolen from its housing in a teletherapy machine at an abandoned clinic in Goiania, in the state of Goias, Brazil. The two people who took the assembly tried to dismantle it and in their attempt the source capsule was ruptured. The remnants of the assembly were sold for scrap to a junkyard owner. He noticed that the source material glowed blue in the dark. Several persons were fascinated by this and over a period of days friends and relatives came to saw the phenomenon. Fragments of the source the size of rice grains were distributed to several families. A few days later, a number of people were showing gastrointestinal symptoms arising from their exposure to radiation from the source.

On 28 September, a doctor in Goiania recognized the characteristic symptoms of radiation overexposure. An emergency response centre was set up and more than 112 000 people were screened for possible contamination; 249 people were found to be contaminated; 20 were hospitalized, and 4 of them died.

A survey of 67 km^3 of the Goiania area showed that eight locations were contaminated. In total, 85 houses were found to have significant contamination, and 200 individuals were evacuated from 41 of them. Decontamination of the affected sites continued until the end of December 1987. As a result, some 3500 m^3 of radioactive waste had to be stored at a temporary site 20 km from Goiania.

Source (67)

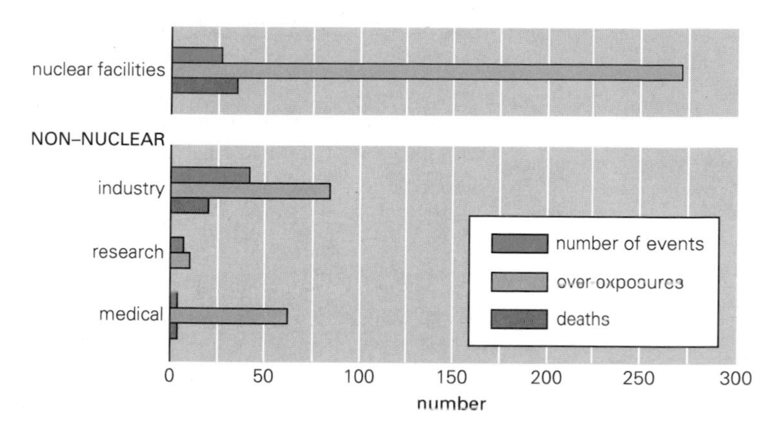

Figure 9.4
Serious radiation accidents reported, 1945–87

based on data from (67)

item in 1984 by the Associated Press editors, and the Ethiopian drought ranked third (63). This natural predisposition towards the dramatic ensures that the information provided by the media about risks is frequently inadequate. . An analysis of 952 print and broadcast news stories in the United States about the Bhopal disaster in the two months following the accident revealed that both print and broadcast reports were event-centred and included little or no discussion of the underlying social, cultural and economic forces that accounted for the chemical plant's construction in India—the main reason for the serious negative impacts of the accident. Instead, news reports focused on the disaster itself, the immediate aftermath, and what was being done to clean up the mess (64). Television news coverage of the Bhopal disaster was the most event-centred. Television did not construe Bhopal in a larger framework of technological hazard, and there was little discussion of the various long-term health, environmental, social or legal issues that the tragedy raised. Only when the news events surrounding the accident subsided did the media pay more attention to such factors (65). The media coverage of the Three Mile Island and Chernobyl accidents has also been found inadequate (66). On the other hand, news coverage of environmental disasters can trigger regional or international action. The coverage of famine and drought in Africa in 1984 brought the crisis to mass public attention, and stimulated public concern and pressure to bring about long-delayed international aid.

The Bhopal accident was ranked as the second biggest news item in 1984 by the Associated Press editors, and the Ethiopian drought ranked third.

The public perception of environmental hazards is of great importance. If people's perceptions are faulty, efforts at public and environmental protection are likely to be unsuccessful. For example, people devastated by a natural disaster often refuse to leave their homesite. And when they are forced to be relocated, they return back as soon as conditions permit (68). Although this has been described as irrational human behaviour, it illustrates how deep-rooted perceptions are difficult to change.

The complexity of public perception is best illustrated by the case of Bangladesh. It is easier to agree on an aid package to build embankments that may hold back the flood waters, than to understand and solve the question of why it is that so many millions of Bangladeshis continue to live on the islands of that country's coastal delta, on the permanent brink of disaster (69).

The International Decade for Natural Disaster Reduction (IDNDR), launched by the General Assembly of the United Nations on the 1 January 1990, aims at instituting an integrated approach to disasters by stimulating the acquisition of data which can then be used in more widespread forecasting and warning systems, by making improvements in disaster preparedness and by changing the sometimes fatalistic attitudes to disasters. Increased community participation, and more and better education and training, will be very important components of the Decade. Achievements of these aims will cause a change in the basic approach to disasters, from the present concentration on post-disaster relief to a future emphasis on pre-disaster preparedness.

The coverage of famine and drought in Africa in 1984 brought the crisis to mass public attention, and stimulated public concern and pressure to bring about long-delayed international aid.

Chapter 10

Toxic chemicals and hazardous wastes

Worldwide, some 10 million chemical compounds have been synthesized in laboratories since the beginning of this century. Approximately one percent of these organic and inorganic compounds are produced commercially—the European Inventory of Existing Commercial Chemical Substances (EINECS) lists 110 000; and 1000–2000 new chemicals appear each year. Some are used directly, as pesticides and fertilizers, but most are intermediate chemicals used for the manufacture of millions of different end products. There is virtually no area of human activity in which chemical products are not used, and many have benefited man and the environment.

But in recent years there has been growing global concern about the harmful effects of chemicals on human health and the environment. The adverse effects of pesticides, vinyl chloride and polychlorinated biphenyls (PCBs) have been well documented since the late 1960s. Over the past two decades many other compounds, such as dioxin, methyl isocyanate (MIC), lead, mercury, other heavy metals and chlorofluorocarbons (CFCs), have become well-known for their effects on health and the environment.

Toxic chemicals

All chemicals are toxic to some degree. The health risk posed by a chemical depends mainly on its toxicity and the length and intensity of exposure to it. Only a few parts per billion of a toxic compound such as dioxin may be sufficient to damage health on brief exposure. In contrast, high doses of other compounds, such as iron oxide and magnesium carbonate, cause problems only after extended exposure.

Over the past two decades there has been an important shift in focus to include the chronic ill-effects of chemicals as well as their

Figure 10.1
Percentage of chemicals with toxicity data

based on data from (1)

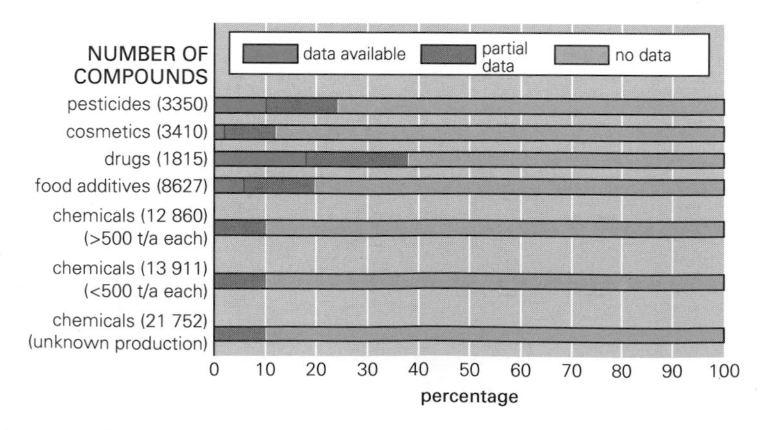

The US National Research Council found that information for a complete health hazard assessment exists for less than 2 per cent of the chemicals produced commercially; for only 14 per cent is there sufficient information to support even a partial hazard assessment.

acute health effects. Chronic effects include birth defects, cancer, genetic problems and neurological disorders. These are of particular concern to the public, and this makes regulatory decisions both more visible and more difficult (see Chapter 18).

Adequate regulation is further complicated by the fact that most chemicals have not been tested sufficiently to determine their toxicity. A study by the US National Research Council (1) found that information for a complete health hazard assessment exists for less than 2 percent of the chemicals produced commercially; for only 14 percent is there sufficient information to support even a partial hazard assessment (see Figure 10.1). Recently, the OECD announced plans to investigate almost 1500 chemicals, each produced in quantities exceeding 1000 tonnes per year, which account for 95 per cent of all chemicals used globally. There is little or no information on their toxicity and virtually nothing is known about their impact on the environment. The first study will include 147 compounds about which there is no toxicological information; 70 of these are produced in quantities exceeding 10 000 tonnes annually each (2). Information has been gathered on the properties of various toxic chemicals and studies are under way to assess their toxicity and hazardous effects (see boxes on page 108 and below).

Toxic chemicals can be released into the environment directly, as a result of human action such as the use of pesticides, fertilizers and solvents; or indirectly, as the waste products of human activities such as mining, industrial processes, incineration and fuel combustion (for accidental releases, see Chapter 9). Chemicals may be released in solid, liquid or gaseous forms, in the air, in water or on land.

The International Register of Potentially Toxic Chemicals

Adequate information for assessing the potential hazards posed by chemicals to human health and to the environment is a prerequisite for their safe use and disposal. In 1976, UNEP established the International Register of Potentially Toxic Chemicals (IRPTC) which collects and disseminates information on hazardous chemicals, including national laws and regulations controlling their use. IRPTC operates through a network of national and international organizations, industries and external contractors, and national correspondents for information exchange which have now been appointed in 112 countries. IRPTC's computerized central data files contain profiles for more than 800 chemicals. In addition, special files are available on waste management and disposal, on chemicals currently being tested for toxic effects and on national regulations covering more than 8000 substances.

The distribution and fate of chemicals in the environment is highly complex and depends on the physical-chemical properties of both the chemicals and the environment. Many chemicals are transported locally, regionally or globally to cause widespread environmental contamination. In California, for example, 16 pesticides and products derived from them were found recently in fog far from where the pesticides were originally used (3). PCBs have been transported in the atmosphere from where they were released in industrial countries to as far as the Arctic. Primarily as a result of consuming contaminated fish and aquatic mammals, Arctic inhabitants are experiencing near-toxic levels of PCB exposure (4). Other examples of transboundary distribution of toxic chemicals have included DDT, mercury, lead and other metals, and hexachlorocyclohexane (5). Other major problems arising from growing global chemical pollution are the effects of chlorofluorocarbons and other chemicals on the ozone layer (see Chapter 2) and of greenhouse gases on climate (see Chapter 3).

The International Programme on Chemical Safety

In 1980, WH0, UNEP and ILO set up the International Programme on Chemical Safety (IPCS) to assess the risks that specific chemicals pose to human health and the environment.

IPCS publishes its evaluations in four forms: as detailed *Environmental Health Criteria* for scientific experts; as short, non-technical *Health and Safety Guides* for administrators, managers and decision makers; as international *Chemical Safety Cards* for ready reference in the work place; and as *Poisons Information Monographs* for medical use.

Responses

To protect human health and the environment, a number of industrialized countries have enacted legislation designed to ensure that industrial chemicals are properly handled and used prior to marketing. However, the task has been complex and slow because the tools for evaluating the effects of chemicals, especially long-term toxicity and environmental hazards, are not sufficiently developed. The assessment of risk to humans based on data from laboratory animals remains controversial, and there are many uncertainties about the methods used to determine potential environmental threats from chemicals. These difficulties have led to maximum levels of exposure being set for some chemicals; to the banning or restricted use of certain substances as too dangerous for marketing and consumption; and to the search for less environmentally harmful substitutes for certain chemicals.

Most developing countries, by contrast, have no toxic chemical control laws, nor the technical and institutional capacity to implement them. There have been several cases in recent years in which products that were banned or severely restricted in industrialized countries have been sold to or dumped on developing countries.

In 1989, UNEP's Governing Council adopted the amended London Guidelines for the Exchange of Information on Chemicals in International Trade, which included a procedure for prior informed consent (PIC). By 1990, 75 countries had nominated national authorities to act as channels for PIC. As a start, PIC has been applied to chemicals banned or severely restricted by 10 or more countries; next it will be applied to those banned or severely restricted by 5 or more. IRPTC notifies participating countries of these bans and offers guidance and training on possible action. Countries then decide whether they wish to ban or allow future imports of the chemicals concerned, and IRPTC channels this information back to exporting countries. It is then up to participating countries to enforce these decisions. Other international legal instruments on the management of chemicals include the FAO International Code of Conduct on the Distribution and Use of Pesticides (amended, 1989); the ILO Convention Concerning Safety in the Use of Chemicals at

Most developing countries have no toxic chemical control laws ... There have been several cases in recent years in which products that were banned or severely restricted in industrialized countries have been sold to or dumped on developing countries.

Work (1990); the OECD recommendations and guiding principles on information exchange related to the export of banned or severely restricted chemicals (1984–85); and the EEC regulation concerning export from and import into the Community of certain dangerous chemicals (1988).

Hazardous wastes

Wastes include substances and objects that are disposed of, are intended to be disposed of, or are required to be disposed of under national law. Certain wastes produced by human activity have been described as hazardous, the term having different connotations in different countries. Wastes containing metallic compounds, halogenated organic solvents, organohalogen compounds, acids, asbestos, organophosphorus compounds, organic cyanides, phenols or ethers are generally considered hazardous (see annexes to the Basel Convention on the Control of Transboundary Movements of Hazardous Wastes and their Disposal for a full list).

Most hazardous wastes are produced by large industries, but hundreds of thousands of small-quantity hazardous waste producers exist, each generating up to 1000 kg of waste a month. They include households, medical facilities (generating biomedical wastes), garages, vehicle-repair workshops, petrol stations, small-scale industries and small businesses. In the United States, 115 000 small-scale hazardous waste producers are now being regulated under the Resource Conservation and Recovery Act (RCRA) and the Hazardous and Solid Waste Amendments (HSWA) (6).

An estimated 338 million tonnes of hazardous wastes are

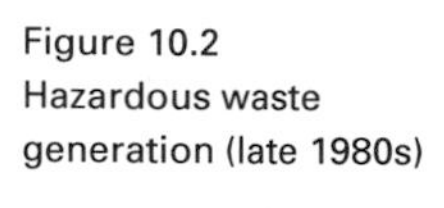

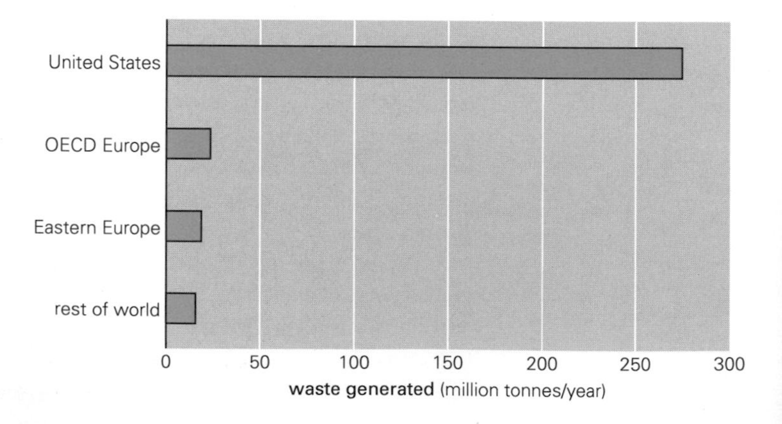

Figure 10.2 Hazardous waste generation (late 1980s)

based on data from (7)

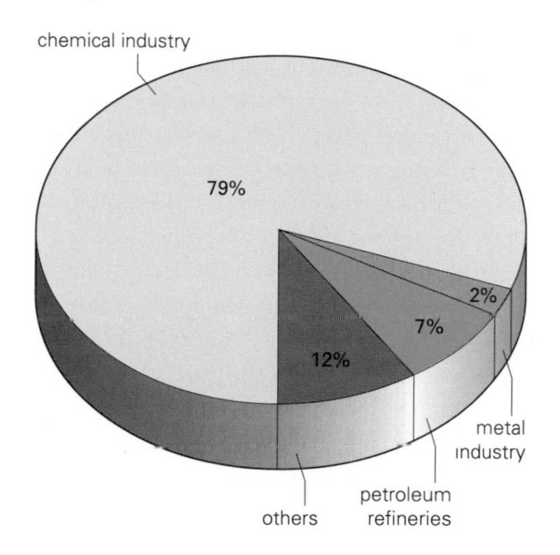

Figure 10.3
Sources of hazardous wastes generated in the United States

based on data from (6)

produced annually worldwide (see Figure 10.2). Of this, 275 million tonnes (81 per cent) are produced in the United States alone (7). By comparison, annual hazardous waste generation in Singapore is 28 000 tonnes, in Malaysia 417 000 tonnes and in Thailand 22 000 tonnes (8).

These figures represent conservative estimates because many countries keep no record of amounts of wastes generated—particularly by small-scale producers. Variations in the composition of these wastes further complicates record-keeping—constituents considered hazardous in some countries may not be considered so in others. In general, the bulk of hazardous wastes comprises chemicals (see Figure 10.3). In OECD Europe the main hazardous waste constituents include solvents, waste paint, heavy metals, acids and oils.

Traditional low-cost methods of hazardous waste disposal include landfill, storage in surface impoundments, and deep-well injection (see Figure 10.4). Thousands of landfill sites and surface impoundments used for dumping hazardous wastes have been found to be unsatisfactory. Corrosive acids, persistent organic compounds and toxic metals have accumulated in these sites for decades. In the Clark Fork Mining Complex in western Montana—the largest site identified in the United States, and considered the largest hazardous waste dump in the world (9)—ponds of wastes from copper and silver

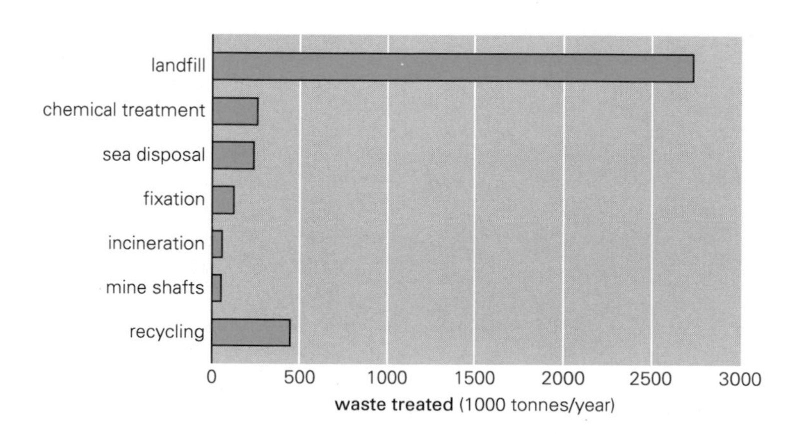

Figure 10.4
Chemical waste management in the United Kingdom (1985)

based on data from (11)

mining and smelting activities have been accumulating for 125 years. When such sites were established little thought was given to their environmental impacts. But when leaks occurred, contaminating groundwater and soil, and threatening public health, policy makers took remedial action, under growing public pressure (see box below). By 1990, the US Environmental Protection Agency had identified 32 000 potentially hazardous sites, about 1200 of which needed immediate remedial action (10). In Europe, 4000 unsatisfactory sites have been identified in The Netherlands, 3200 in Denmark and some 50 000 in western Germany (7). Although some industrialized countries have begun to clean up the problem sites, the cost of remedial action to counteract long years of neglect has been found to be very high. Estimates indicate that about $30 billion are needed for remedial operations in western Germany, $6 billion in The Netherlands, and about $100 billion in the United States (7).

Unsatisfactory dumping of hazardous wastes has also exposed people directly to hazardous chemicals. Perhaps the most notorious incident was the outbreak of Minamata disease in Japan in the 1950s and 1960s, as a result of discharges from a chemical factory into the sea that contaminated fish with mercury. The fish were eaten by local people from the town of Minamata on Kyushu Island, Japan, and thousands suffered neurological disorders. This and a similar incident at Niigata on the east coast of Honshu caused the death of about 400 people. Although dumping waste at sea is controlled under international and regional conventions, several countries are still disposing of hazardous wastes in this way. About 10–15 per cent of

The Love Canal

From 1942 to 1953 a chemical company disposed of some 21,800 tonnes of chemical wastes in a trench in the city of Niagara Falls, New York, which contained the remnants of an old canal—the Love Canal.

Shortly after the company ceased its landfilling in 1953, a school and several buildings were constructed at the site. Heavy rains in the winter of 1975 and spring of 1976 caused land subsidence and created ponds of surface water that were heavily contaminated with chemicals from the dump. The contaminated waters infiltrated into nearby residences, causing public concern and complaints about possible health hazards.

In August 1978 an emergency programme was undertaken to relocate the residents of 238 houses in the area. About $100 million has been spent on site remediation, resident relocation and investigations at the Love Canal.

Source (10,12)

hazardous wastes produced in Europe are dumped at sea (7).

In the early 1980s, the problem of transfrontier hazardous waste movement became an important issue in Europe and in North America—particularly after a well-publicized incident in which a consignment of drums containing dioxin-contaminated mud disappeared mysteriously in transit between Italy and France. One reason for the transfrontier movements of waste may be that legal disposal in a foreign country is less expensive than at home and that there is no disposal capacity for these wastes in the country of origin. On average, a consignment of hazardous wastes crosses an OECD European frontier every five minutes, and there are more than 100 000 such movements in OECD European countries each year. Between 2.0 and 2.5 million tonnes of hazardous wastes crossed OECD European frontiers in 1988 (13). North American figures available indicate that about 230 000 tonnes of hazardous wastes were exported and about 9000 crossings were made in the same year (7). Hazardous wastes have also been legally moved between OECD and non-OECD countries, including about 200 000–300 000 tonnes of hazardous wastes transported annually from EC to East European countries. North America and European countries have also exported wastes to developing countries, and Europe sends about 120 000 tonnes of hazardous wastes to the Third World each year (2).

In 1985, the OECD adopted a number of principles to control the transfrontier movements of hazardous wastes. These have been embodied in EEC law and were endorsed by the OECD in 1988, when a core list of hazardous wastes and other wastes that should be controlled in transfrontier movements was established. As controls over the movements of hazardous wastes and their disposal in industrialized countries have been tightened, so the illegal dumping and movement of wastes has increased. Of particular concern have been the shady deals involving dumping hazardous wastes in developing countries. Africa, the Caribbean and Latin America have been improperly used as disposal sites for a wide range of wastes from the industrialized world. Illegal traffic and dumping of hazardous wastes have also been reported in Asia and the South Pacific (14) and even in Europe.

On average, a consignment of hazardous wastes crosses an OECD European frontier every five minutes, and there are more than 100,000 such movements in European countries each year.

Responses

Growing international concern over the transfrontier movement and dumping of hazardous wastes, especially in developing countries, led to the adoption in 1989 of the Basel Convention on the Control of Transboundary Movements of Hazardous Wastes and their Disposal (see box below). The complexities of managing wastes in general and hazardous wastes in particular have intensified with the realization that old, uncontrolled landfills containing hazardous wastes pose serious environmental risks; with the discovery that illicit international trafficking in hazardous wastes was taking place; and with the growing reluctance of the public at large to accept landfills or treatment plants in their neighbourhood—known as the not-in-my-backyard syndrome (NIMBY).

Although above-ground storage and 'controlled' burial of waste are the most common methods of hazardous waste management, some countries, such as Denmark, Finland, The Netherlands and the United States, plan to ban landfills that use no form of waste pretreatment. There is a growing tendency to use specific pretreatment techniques for certain wastes. In Austria, Germany and Switzerland, all hazardous liquid organic wastes should be incinerated or subjected to physical or chemical treatment. Incineration, especially high-temperature incineration using plasma arc furnaces, is increasingly used to treat hazardous wastes.

Waste reduction or prevention is certainly the best way to protect human health and the environment. Given the cost and

The Basel Convention

The Basel Convention on the Control of Transboundary Movements of Hazardous Wastes and their Disposal was adopted by 116 governments and the European Community on 22 March 1989. The ultimate aim of the Convention is to reduce the generation of hazardous wastes to a minimum. Its current targets are to control the permitted transboundary movements and disposal of hazardous wastes.

The Convention outlined the general obligations of states in relation to the transboundary movements of hazardous wastes, defined illegal traffic in hazardous or other wastes, and outlined the responsibilities of the parties involved; it also enumerated the principles of international cooperation that would improve and achieve environmentally sound management of hazardous and other wastes.

As of 31 December 1990, 52 countries and the European Community had signed the Basel Convention, and five countries had ratified it.

More vigorous research and development in waste minimization and recycling ... could probably cut the production of hazardous wastes in many industrialized countries by a third by the year 2000.

difficulty of handling waste, the principle of pollution prevention pays (the three Ps) should be widely promoted (see box opposite). Its benefits are enormous. Occupational and public exposure to hazardous chemicals is reduced; industrial efficiency and competitiveness are enhanced as waste prevention simultaneously cuts raw material input, saves energy, and reduces the volume of waste to be stored, treated, or disposed of; less waste means less expenditure on buying and operating pollution control equipment; accidents during rail and highway transportation of waste are reduced; the need for off-site hazardous waste facilities are reduced, which also reduces associated health, environmental and political problems. Companies can reduce liability risks and costs otherwise arising from inadequate disposal practices—in fact up to 50 per cent of all environmental pollutants and hazardous wastes could be eliminated with existing technology (15).

Waste has been recycled and reused in some countries for decades for economic reasons—the best known examples being the reuse of scrap metals and the reuse of glass bottles for soft drinks. Recycling is now receiving increased attention in many countries—for example, in Hungary about 29 per cent of hazardous wastes are being recycled (16). There is a great potential for recovering certain materials, such as solvents and chromium, copper, mercury and other metals. It has been estimated that up to 80 per cent of waste solvents and 50 per cent of the metals in liquid waste streams in the United

Pollution prevention pays

The following examples illustrate the economic feasibility and environmental benefits of reducing or preventing waste—in other words that pollution prevention pays:

- The 3M Company (Minnesota Mining and Manufacturing) executed a 3P programme covering more than 2000 projects. The programme saved the company $420 million over 10 years and prevented the annual discharge of 12 000 tonnes of air pollutants, 14 000 tonnes of water pollutants, and 313 000 tonnes of sludge and solid waste.
- Exxon Chemical Americas installed 16 floating roofs on open tanks of volatile chemicals at its Bayway plant. This resulted in annual savings of 340 tonnes of organic chemicals, worth about $200 000 in addition to a marked reduction of releases into the environment.

Sunkiss, a French Company, developed a low-emission paint-drying technique. The process reduces the emission of evaporated solvents by 99 per cent by destroying them in the heating/drying process; it also reduces drying time by 99 per cent, and reduces energy use for drying by 80 per cent. Energy saving alone recovers the cost of the device in two months.

Source (15)

States can be recovered using existing technologies (17). In Japan, the United States and western Europe, waste exchanges, operating on the simple premise that one industry's waste can be another's raw material, have succeeded in promoting the recycling and reuse of industrial waste. Most serve as information clearing-houses, publishing catalogues of wastes available and wastes wanted, to inform industries of trading opportunities. A successful recycling trade benefits both buyers and sellers; the buyers reduce their raw material costs, the sellers their treatment and disposal costs.

Several technologies are available for dealing with the hazardous wastes generated by industry. But more vigorous research and development in waste minimization and recycling, together with increased technical and financial support to encourage investment—and in some cases a waste generation tax—could probably cut the production of hazardous wastes in many industrialized countries by a third by the year 2000.

In Japan, the United States and western Europe, waste exchanges, operating on the simple premise that one industry's waste can be another's raw material, have succeeded in promoting the recycling and reuse of industrial waste.

Part II

Development Activities and Environment

Chapter 11

Agriculture and food production

At the beginning of the 1990s, the average per capita food consumption was 2670 calories—a level considered nutritionally adequate. However, this global average has little significance because food consumption is inadequate in many developing countries. There is a gap of 965 calories per capita between the developed and the developing countries (3399 and 2434 calories per capita, respectively), and there are wide gaps between and within the developing countries themselves (1). In fact, the increase in per capita food availability in the developing countries as a whole slowed in the 1980s compared with the 1970s and 1960s; the situation for some countries, for example those in sub-Saharan Africa, worsened so much that per capita food availability in 1989 was less than in 1970 (2).

This worldwide disparity has been created and aggravated by a combination of social, economic, environmental and political factors, including a fall in commodity prices, agricultural subsidies in developed countries, agricultural trade barriers, inequitable access to resources and products, and the often primitive conditions of production and processing of agricultural output in many areas. As a result, the number of chronically hungry people in the world increased from about 460 million in 1970 to about 550 million in 1990, and is expected to reach 600–650 million by the year 2000 (2). Close to 60 per cent of the hungry people in the developing world live in Asia, about 25 per cent in sub-Saharan Africa and some 10 per cent in Latin America and the Caribbean.

The fact that hunger is closely related to poverty is well established. According to the World Bank (3), 1116 million people in the developing countries are living in poverty, and 630 million of them

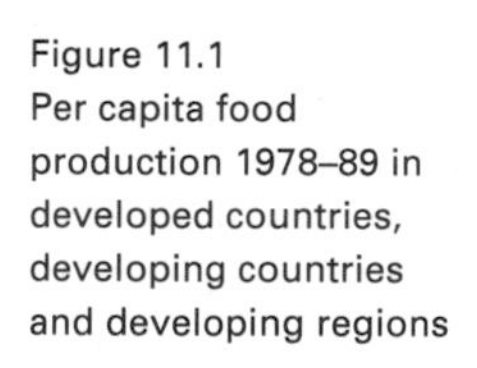

Figure 11.1 Per capita food production 1978–89 in developed countries, developing countries and developing regions

based on data from (12)

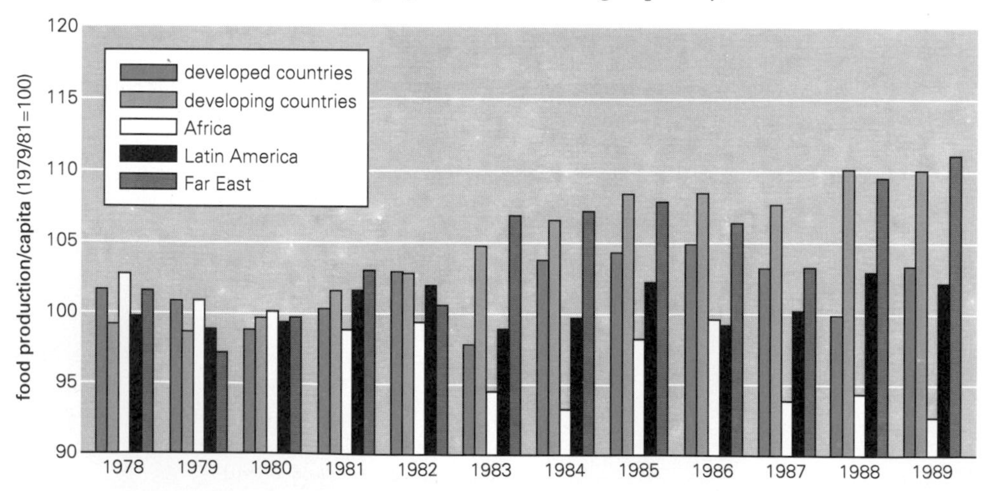

The increase in per capita food availability in the developing countries slowed in the 1980s compared with the 1970s and 1960s; the situation for some countries ... worsened so much that per capita food availability in 1989 was less than in 1970.

can be considered extremely poor. This last group is the most threatened by hunger and chronic malnutrition.

Agricultural output and food production increased in both developed and developing countries in the period 1970–90. The annual rate of increase was higher in the developing countries (about 3.0 per cent) than in the developed countries (about 2.0 per cent). In the latter, there was a near stagnation in per capita food production in the 1980s, with marked drops in 1983 and 1988 due to unfavourable weather conditions, particularly in North America. In the developing countries there were major increases in Asia, a near stagnation in Latin America and a marked drop in Africa (Figure 11.1). The rate of increase in cereals production (Figure 11.2) was higher in the developed countries than in the developing countries (about 32 per cent and 15 per cent, respectively, between 1970 and 1990). In the developed countries, the annual rate of cereals production was higher than population growth (about twice as much), but in the developing countries it was much lower (about one-fifth). A wide gap, currently 529 kg per capita, continues to exist between the annual cereals output of developed and developing countries as a whole (777 kg/capita and 248 kg/capita, respectively, in 1990) (4).

About 12 per cent of the world's population is entirely dependent on livestock production. On average, one-quarter of the gross value of agricultural production is attributed to livestock production. When considering the non-monetized contribution of livestock (through the provision of draught power and manure), this

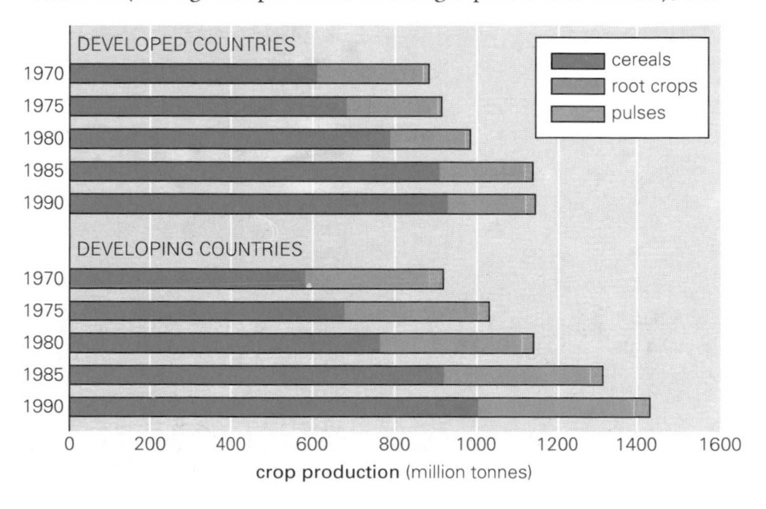

Figure 11.2
Crop production in developed and developing countries, 1970–90

based on data from (1, 31)

percentage amounts to 44 per cent (5). The largest share of the world's livestock population is found in the developing countries—99.5 per cent of buffaloes, 98.5 per cent of camels, 94.0 per cent of goats, 68.5 per cent of cattle, 57.8 per cent of pigs, and 52.5 per cent of sheep (1989 figures). However, meat production in developing countries is much lower than in developed ones (68.7 and 103.2 million tonnes, respectively, in 1990) (4). This is mainly because most livestock in the developing countries are raised in traditional, small-scale farming systems, primarily for subsistence purposes, with additional income being generated from the sale of animal products. Animals also supply power for agriculture; in Asia and Africa, they provide 28 and 10 per cent of agricultural power.

Fisheries produce 16 per cent of the animal protein available in the world—roughly the same as that provided by beef and by pork (6). Most of the world's fish production comes from the marine areas (Figure 11.3), which accounted for about 86 per cent of production in 1990. Of this, 90 per cent is estimated to be from coastal areas. About 14 per cent comes from inland (fresh) waters. Of the total catch, about 7 million tonnes come from freshwater aquaculture and about 5 million tonnes from mariculture. In other words, about 11 per cent of global fish production comes from aquaculture. At present growth rates, aquaculture production should almost double by the end of the century (6).

Most aquaculture is in Asia, where production amounts to about 4 million tonnes per year (7). Coastal aquaculture (shrimp) in Asia accounted for 82 per cent of world cultured shrimp in 1990 (about

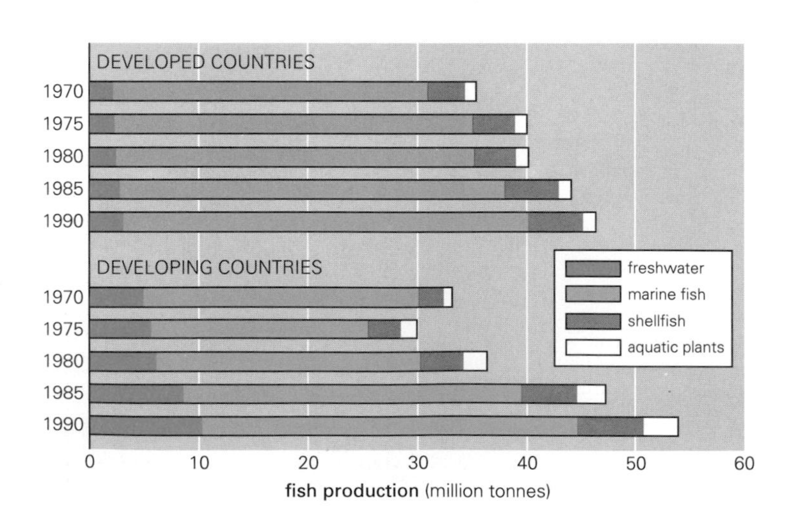

Figure 11.3
Fish production in developed and developing countries, 1970–90

based on data from (1, 6, 31)

If ... no new land is brought under cultivation and no existing land goes out of production ... the arable land available per head of population will decline to 0.23 ha in 2000, to 0.15 ha in 2050 and to 0.14 ha in the year 2100.

400 000 of the 471 000 tonne world total). Most freshwater aquaculture in Asia is for local consumption in rural areas. Small-scale systems such as rice-fish culture and integration of aquaculture with livestock are common in many Asian countries.

Agriculture, resources and environment

The total area of potential arable land in the world is about 3200 million hectares, about 46 per cent of which (1475 million ha) is already under cultivation. Worldwide, the area of arable land increased by only 4.8 per cent over the period 1970–90; the increase in developed countries was 0.3 per cent and that in the developing countries was 9 per cent (Figure 11.4). The area of arable land available per head of population, however, fell from 0.38 ha in 1970 to 0.28 ha in 1990 mainly as a result of population growth and loss of land to agriculture (Chapter 6). The decrease was most noticeable in the developing countries where it fell by 29 per cent (from 0.28 to 0.20 ha/capita). In the developed countries, the decrease was 12.5 per cent (from 0.64 to 0.56 ha/capita). It has been estimated that if the arable land area is maintained at the present level (1475 million ha), no new land is brought under cultivation and no existing land goes out of production as a result of degradation, the arable land available per head of population will decline to 0.23 ha in 2000, to 0.15 ha in 2050 and to 0.14 ha in the year 2100 (8).

Figure 11.4
Arable and permanent cropland, and arable land per capita, 1970–90

based on data from (12, 31)

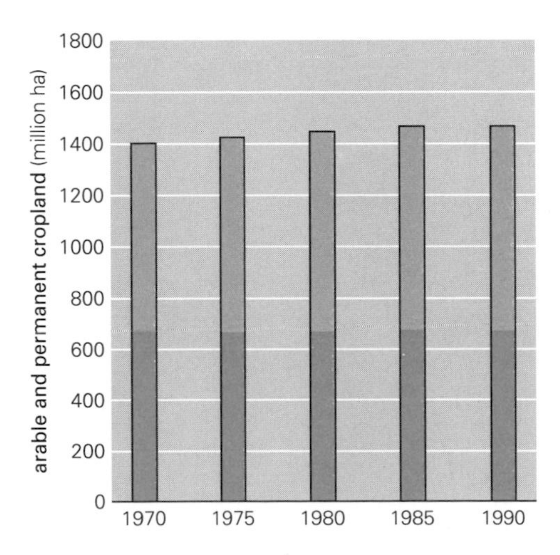

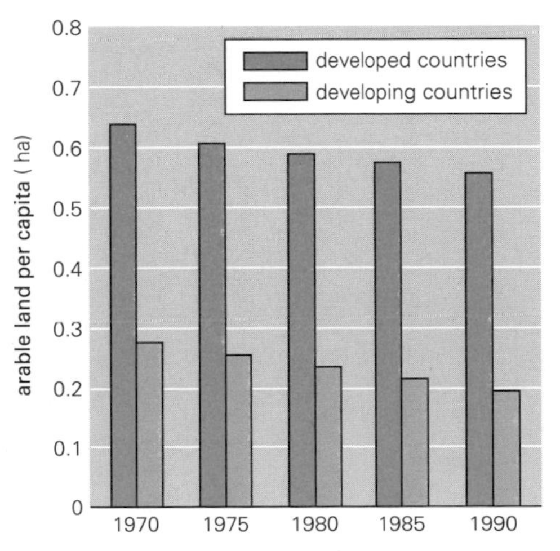

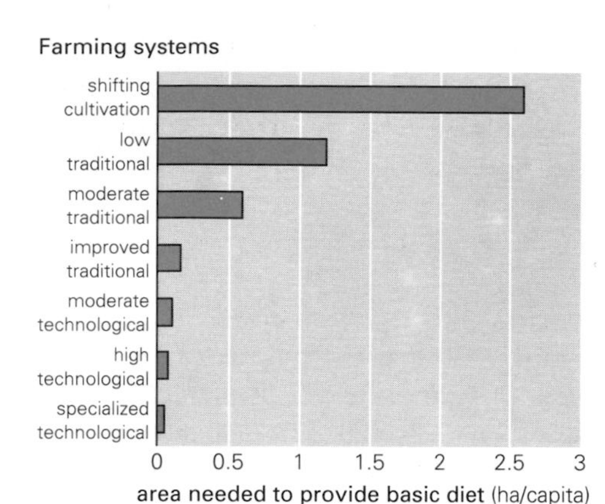

Figure 11.5
Arable land needed to meet basic food need

based on data from (11)

It has been claimed that large areas of new land could be brought under cultivation (9, 10). But unused arable land is not always available to those who need it most, and opening up new areas remains an expensive means of increasing agricultural production. In fact, further expansion of agricultural land is constrained in many parts of the world. In tropical Africa, for example, agricultural and livestock development is severely hindered because of such diseases as river blindness (onchocerciasis) and human and animal trypanosomiasis. The latter renders livestock production virtually impossible over some 10 million km^2 of high rainfall areas—45 per cent of all the land in sub-Saharan Africa. In arid regions, shortage of water for irrigation constitutes a major constraint on future expansion of the cropland area.

It has been estimated that, with traditional agriculture, the minimum diet for one person can be provided from an average of 0.6 ha of arable land (11). This means that the area now under cultivation could provide the minimum diet for less than half the world population. There has therefore been no alternative but to increase the output of existing arable land through technological innovations. Efforts to do so have been successful; productivity gains have been achieved largely by using the 'green revolution' technological packages that require the use of high-yielding varieties (HYVs) of seeds and high inputs of water, fertilizers and pesticides. This has led to a drop in the average per capita land requirement to meet basic needs, and the application of more advanced technologies (such as

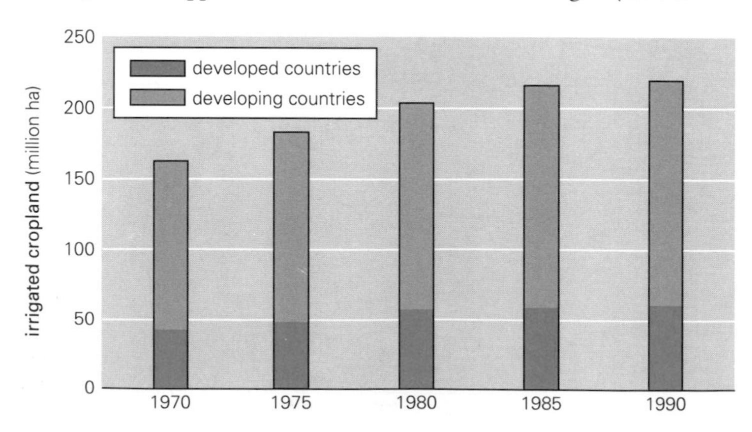

Figure 11.6
Irrigated cropland in developed and developing countries, 1970–90

based on data from (12, 31)

biotechnologies) would lead to a further decrease (Figure 11.5). However, the intensification of agriculture requires high inputs, and the more impoverished an ecosystem, the more inputs are needed to raise outputs. This has implications for resource use and for the state of the environment.

Worldwide, about 2700 km^3 of water were withdrawn for irrigation in 1990, or about 69 per cent of all the freshwater used (Chapter 5). The world's irrigated land increased from 168 million ha in 1970 to 228 million ha in 1990 (5, 12), an increase of about 36 per cent in two decades (Figure 11.6). Although the land now under irrigation accounts for one-sixth of the cultivated land, it produces one-third of the world's food (which means it has more than twice the productivity of average rainfed land). However, the rate of expansion of irrigated land has been slow because the availability of additional irrigable land and good quality water are severely constrained in many parts of the world. The scarcity of water resources is compounded by the loss of irrigation water in supply systems and on the farms. Such losses are generally in the range of 50–60 per cent, but may reach as much as 75 per cent in some countries (13). In most developing countries, irrigation water is supplied free or is heavily subsidised (13, 14). This has led to inefficient use of water for irrigation and has discouraged simple conservation measures, which are more likely to be carried out if farmers pay for irrigation water. Studies have demonstrated that each 10 per cent rise in the price of water generates about 6 per cent savings in water use (14).

The increased application of chemical fertilizers supplying plant nutrients (nitrogen, phosphorus and potassium) is an essential component of modern agriculture. World consumption of chemical

Figure 11.7
Fertilizer use, and fertilizer use per hectare, in developed and developing countries, 1970–90

based on data from (32)

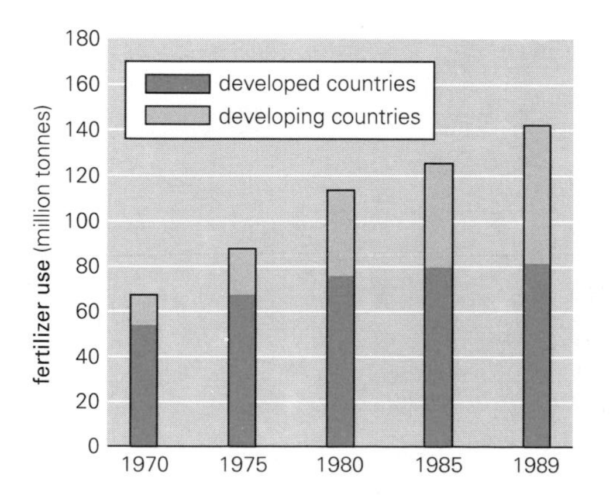

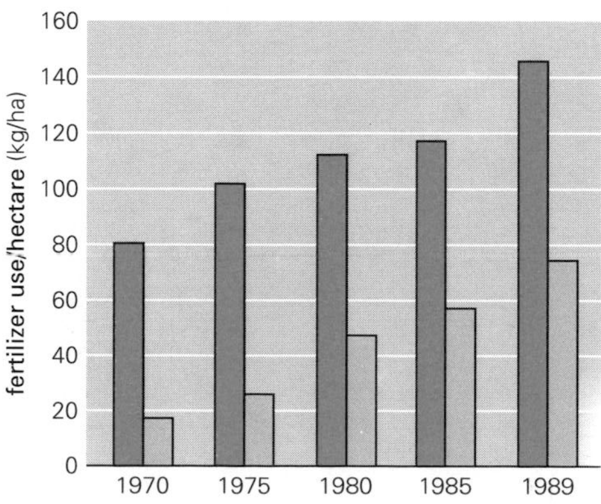

fertilizers more than doubled over the past two decades, from about 69 million tonnes in 1970 to about 146 million tonnes in 1990 (Figure 11.7). The rate of increase was much higher in the developing countries (360 per cent) than in the developed ones (61 per cent). Nitrogenous fertilizers are the most heavily used, followed by phosphates and potash. Fertilizers have been used much more intensively in the developed countries than in the developing ones, although the rate of application in the latter has been rising fast (by 327 per cent during 1970–89), as a result of the introduction of green revolution packages. About 50 per cent of the fertilizer used benefits the plants; the remainder is lost from the soil system by leaching, run-off and volatilization (15). Fertilizer subsidies in many developing countries have led to inefficient application, with consequent economic losses and increased environmental damage on and off farms.

Crops are affected by both pests and weeds. In North America, Europe and Japan crop losses caused by pests are estimated at 10–30 per cent. In the developing countries they are often 40 per cent, and losses of 75 per cent have been reported, for example for maize in Africa. Pests affect not only the yield of food and feed crops but also their quality.

About 90 per cent of pesticides sold are used in agriculture; the remainder is used in public health programmes (16). The growth of world pesticide use is normally measured in sales because information on the weight and volume of the active ingredients produced is scarce. It has been estimated that sales of pesticides increased from US$7700 million in 1972 to US$15 900 million in 1985, and reached about US$25 000 million in 1990 ($1985); the major pesticides are herbicides (46 per cent), insecticides (31 per cent) and

Figure 11.8
Numbers of pests resistant to pesticides, 1908–88

based on data from (33)

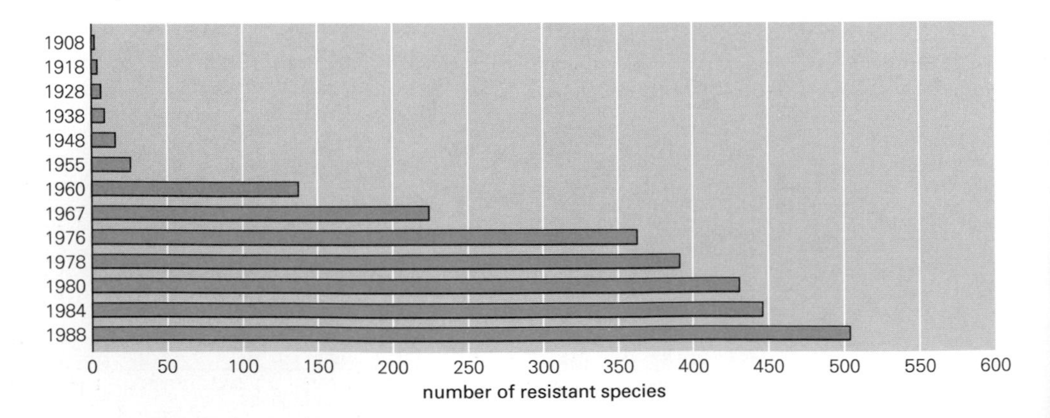

fungicides (18 per cent). About 80 per cent are used in the developed countries. However, the growth rate in the developing countries (7–8 per cent per year) is larger than in the developed ones (2–4 per cent per year). It has been estimated that more than 90 per cent of pesticides do not reach target pests (17) and contaminate land, water and air. Repeated applications of pesticides (often highly subsidised in developing countries) have led to the build-up of resistance among target pests (Figure 11.8). In several cases this has prompted the use of other, more toxic pesticides, with greater occupational and environmental risks.

Agriculture is a modest user of commercial energy relative to other economic sectors, accounting for an estimated 5 per cent of commercial energy use in the world, or about 375 million tonnes of oil equivalent per year. This estimate takes into account energy used in irrigation, pesticide and fertilizer production, and machinery operation, but not energy used in food processing, storage and transportation. The developed countries—where the use of fertilizers, pesticides and farm mechanization are high—account for about 77 per cent of commercial energy use in agriculture.

Impact of agriculture on the atmosphere

Agriculture contributes both to local and global atmospheric pollution. Shifting, or swidden, cultivation and the burning of savanna lands and the clearing of forest and savanna for livestock and arable farming produce carbon dioxide, carbon monoxide, methane, nitrogen oxides, ammonia, sulphur oxides and particulate matter. The amount of biomass burned between 1970 and 1990 has been estimated at 4.9–8.9 billion tonnes annually, about 60–65 per cent of which is directly related to agriculture, the rest to wildland fires, the burning of industrial and fuelwood and other forms of deforestation (18).

Paddy fields and the guts of livestock produce considerable amounts of methane. Ammonia is released from livestock waste; the largest emissions occur in Argentina, Brazil, China, India and the United States, each producing more than one million tonnes of ammonia-nitrogen annually (18).

Even if nitrogen fertilizers are not used, cultivated soils sometimes emit large amounts of

Ammonia is released from livestock waste; the largest emissions occur in Argentina, Brazil, China, India and the United States, each producing more than one million tonnes of ammonia-nitrogen annually.

Estimates of major emissions into the atmosphere due to agriculture

	million tonnes per year	*per cent of anthropogenic emissions*
carbon dioxide	1200	17
methane	230	66
nitrous oxide	2	71
ammonia	28	80
sulphur oxides	2	2
nitric oxides	5	7
particulate matter	20	35

Emissions include contributions from burning of biomass directly related to agriculture, paddy rice, livestock rearing and use of fertilizers.

Source (26)

nitrous oxide (especially in the tropics), perhaps as much in the aggregate as that released from fertilized fields (19). Application of fertilizers increases nitrous oxide emissions. Air can easily become contaminated with pesticides during spraying operations. Traces of pesticides have been found in fog in California (20) and in rain. It has been shown that even relatively non-volatile pesticides, such as DDT, evaporate into the atmosphere quite rapidly, particularly in hot climates, and can be transported over long distances, contributing to what has been called global chemical pollution. The box summarizes the estimates made for the major emissions into the atmosphere due to present agricultural practices.

Impact of agriculture on water

Excessive irrigation wastes large quantities of water, leaches out soil nutrients and micronutrient trace elements, and creates problems of secondary salinization and alkalinization, which have damaged millions of hectares of productive lands (Chapter 7). Over-exploitation of groundwater for irrigation has led to the depletion of groundwater resources in arid areas such as the Middle East; and, in coastal zones, it has resulted in excessive intrusion of salt water from the sea into groundwater aquifers. In several countries inadequately designed and operated irrigation systems have created ideal breeding conditions for the organisms that cause such water-borne diseases as schistosomiasis, liver fluke infections, filariasis and malaria (Chapter 18). These diseases are not new, but their incidence has increased as new irrigation schemes have been introduced.

Because water of good quality is not always available, there is a growing tendency to use water of marginal quality for irrigation. For example, brackish water is used in some Gulf States in the Middle East, and drainage water mixed with fresh water is used for irrigating some crops in Egypt (13). Unless the use of such water is carefully managed and monitored, it can lead to salinization and deterioration of the quality of groundwater in the aquifers near irrigated lands. Although municipal wastewater has been used for irrigation for centuries, care has always been needed. Pathogenic bacteria, parasites, and viruses are all found in sewage, and may survive treatment processes. Once in the environment, many are able to exist for prolonged periods; cholera and typhoid outbreaks have been closely associated with wastewater irrigation (21).

Despite their low solubility, pesticides can be leached into drainage water, which can then pollute the surface and coastal water into which drainage water is discharged. Pesticides have been detected in coastal waters and in fresh-water bodies in many regions (Chapters 4, 5). Groundwater contamination is also common where pesticides are heavily used. In California, in 1980–84, dibromochloropropane was detected in some 2000 wells over an area of 18 000 km^2. The herbicides atrazine, alachlor and simazine are also important contaminants. Aldicarb, a nematocide, has become a common contaminant of aquifers below potato fields and citrus orchards in several countries (18). The pollution of surface and groundwater with pesticides can affect aquatic life and human health. Pesticides, especially persistent ones, can build-up through the food chain, with consequent risks to humans (Chapter 18).

Fertilizers can be easily leached into drainage water, and when such water is discharged into rivers or the sea, the leached nutrients (nitrogen and phosphorus) create widespread eutrophication. Nitrate and phosphate have been responsible for generating dense algal growth which has harmed fish and other aquatic life (Chapters 4 and 5). In Sweden, in 1989, it was estimated that 26 per cent of the total nitrogen load in the country's surrounding seas came from agriculture, 23 per cent from forests and forestry, 8 per cent from wetlands, 19 per cent from municipal and rural sewage, 4 per cent from industry, 10 per cent from atmospheric deposition, and 10 per cent from other land uses (22).

Fertilizers are not the only source of nutrients. Nitrate-nitrogen is a common pollutant on large feedlots. Wastes from feedlots are becoming a major source of water pollution in several industrialized countries. For example, in England and Wales, 20 per cent of the annual number of pollution incidents recorded by the water authorities in 1988 is

In California, in 1980–84, dibromochloropropane was detected in some 2000 wells over an area of 18 000 km^2.

from feedlot wastes (18). Nitrates from fertilizers and feedlot wastes have caused the contamination of groundwater in many countries, and the issue is of major concern in Europe and in North America. According to WHO (23), water becomes unsuitable for drinking when its nitrate concentration exceeds 45 ppm. The EC has issued a directive which would require any area where the nitrate concentration in surface or groundwater exceeded 50 ppm to be declared a 'vulnerable zone' in which there would be compulsory and automatic restrictions on farming.

Impact of agriculture on land

The pressures to expand the area under cultivation have resulted in more and more utilization of marginal land. Cultivation on steep hillsides and increasing rates of deforestation, especially in the tropics, have led to soil degradation, declines in productivity and desertification (Chapters 6 and 7). The draining of wetlands for conversion to agricultural uses has detrimental effects on fish, wildlife and wetland habitats. Increased use of estuarine areas, the nurseries for most of the coastal fish stock, may affect bay, river mouth and shallow coastal habitats (Chapters 4 and 8). The development of aquaculture often requires the conversion of large areas of coastal lowlands into ponds. For example, in 1980 the coastal area under extensive pond culture amounted to 176 000 ha in the Philippines, to 192 000 ha in Indonesia, to 2500 ha in Thailand, and to 12 000 ha in India (24). In these and other countries—such as Malaysia and Ecuador— large areas of coastal mangroves and marshlands are being converted to ponds. Mangrove swamps act as protective areas between land and sea, and are the habitat of many terrestrial and aquatic organisms. The conversion of mangroves into ponds for aquaculture will not only affect this habitat, but lead to other environmental impacts—for example, the destruction of mangroves removes a natural barrier to the storm surges that accompany cyclones (Chapter 9).

Rangeland has been degraded in many parts of the world as a result of mismanagement and overgrazing (Chapter 6). The situation is especially precarious in arid and semi-arid lands. The grass and shrub vegetation in extensive rangeland areas in the Middle East and North Africa have undergone extensive changes, particularly because of overgrazing (13). Rangeland in the region is particularly prone to xerification, or drying out, that can be caused or accelerated by drought and/or overgrazing.

Impacts of use of high-yielding varieties of seed

The extensive use of HYVs of seeds has led to a marked decrease in genetic diversity. For example, the spread of HYVs of wheat and rice since the mid-1960s has inadvertently caused a loss of the gene pools in such centres of crop diversity as Afghanistan, Iran, Iraq, Pakistan and Turkey (13). In 1980, there were as many as 30 000 varieties of rice in India. By the end of this century, it is estimated that as few as 12 varieties will dominate 75 per cent of that country. In addition, the uniformity of the genetic background of HYVs opens up the possibility of lower resistance to new diseases or pests.

Some of the limitations of HYVs that have received increasing attention in recent years stem from their dependence on the presence of a whole package of complementary inputs such as water, fertilizer, and pesticides which are not always readily available in developing countries. In areas with conditions favourable to the adoption of the new varieties, especially as far as water availability is concerned, the use of the new seeds spreads rapidly. In areas with less favourable conditions,the new varieties offer little or no advantage over traditional farming methods (25).

The use of HYVs has also created several socio-economic problems. Small farmers are generally unable to acquire the HYV packages, and their farm yields remain low. Many have, therefore, been forced to abandon agriculture. On the other hand, increasing numbers of farmers have switched to cultivation of 'urban consumer' or 'export' crops, which are more profitable. This has not only disturbed agricultural structures in some countries, but has also negated the main rationale behind the introduction of packages of HYVs, which is to increase the yields of the main staple food crops (13).

Agricultural residues and livestock waste

Worldwide, farm crops leave substantial residues, the extent and scale of which are rarely realized. The amount of such residues was estimated at about 930 million tonnes in 1970 and about 1500 million tonnes in 1990 (26), about 75 per cent of which was cereal straw and residues from maize and barely crops. These residues must be removed from

In 1980, there were as many as 30 000 varieties of rice in India. By the end of this century, it is estimated that as few as 12 varieties will dominate 75 per cent of that country.

the fields to control pests and diseases, and to prevent fouling of the soil for the next crop. In several countries, especially industrialized ones, most of the residues are burned in the field. In developing countries, however, substantial quantities of these residues are used as fuel, mainly for domestic purposes (Chapter 13), as additives to animal dung to make dung cakes for fuel, or to mud to make mud bricks for building (25). A substantial amount of cereal straw and other residues is also used as animal feed.

Livestock produced about 1500 million tonnes (air-dry) of dung in 1970 and about 2200 million tonnes in 1990 (26). This waste constitutes a major source of pollution, especially in developed countries near animal farms. The pollution this waste causes in the air and in water has already been described. In developing countries dung is widely used as fuel in many rural areas in the form of dung cakes, and to produce biogas for fuel, especially in China, India and other Asian countries (Chapter 13). The residues from biogas plants, which are rich in nutrients, have been used as fertilizer and/or for feeding algae and fish ponds (27).

Responses

Agricultural impacts on the environment can be viewed in the context of a system comprising three inter-related components: agricultural resources, agricultural technology and the environment. The quantity, quality and availability of resources determine the technologies to be used. The technologies employed, in turn, have environmental and/or socio-economic impacts, generating demands for other technologies and/or policies to reduce or eliminate the negative impacts. Agricultural practices that lead to environmental degradation will trigger or exacerbate the neglect of land and of rural development (a symbiosis exists between agricultural and rural development but this is not fully appreciated in some developing countries), leading to increased migration from the countryside to the towns and cities. This will not only aggravate urban problems, but will also undermine efforts to increase indigenous food production, and hence increase national dependence on imported food. It is therefore in the interest of national stability and security that countries develop and implement environmentally sound agricultural development plans.

Much current research is aimed at increasing agricultural productivity in an environmentally sound manner. A number of international and regional organizations are also supporting research and development activities with the same goal. Many activities of FAO, IFAD, UNEP and bodies such as the International Board for Plant Genetic Resources (IBPGR), the Consultative Group on International Agricultural Research (CGIAR), the International Rice Research Institute (IRRI), the International Centre for Maize and Wheat Improvement, the International Centre for Insect Physiology and Ecology (ICIPE) are geared to that goal.

Many simple technologies have been developed to increase the efficiency of use of different inputs in agricultural systems. Adjusting the timing of application of fertilizers and the amounts used has led to a considerable saving of fertilizer, with both economic and environmental benefits (18). The use of sulphur-coated urea (SCU) on rice has led to the controlled release of nitrogen and hence a lowering of the concentration of nitrogen in both the soil and water at any given time. Although costs are more than for ordinary urea, the economic returns are potentially of the order of US$6–7 for every dollar spent, not counting the environmental benefits (28). The recourse to biological processes for fertilization (nitrogen-fixing plants, crop rotations, use of trees as 'nutrient pumps', recycling of wastes) is growing in several countries, especially industrialized ones, where it is sometimes referred to as 'ecological' agriculture. There is also a growing tendency to apply the concept of integrated plant nutrition systems (IPNS) which involves the use of carefully derived combinations of mineral and organic fertilizers which are applied in combination with complementary crop practices such as tillage, rotation and moisture conservation. As a result, soil quality is conserved and pollution reduced to a minimum.

In the past two decades, increased attention has focused on the use of integrated pest management (IPM) to keep pests and diseases at an acceptable level. IPM strategies include the selective use of pesticides and rely on the use of biological methods, genetic resistance and appropriate management practices. Although the application of IPM has been slow, especially in relation to food crops, many success stories have demonstrated its viability (see box on next page). In the United States, IPM is now used on about 15 per cent of the total area of cultivated land (18), and its use is growing, for example, in Central America and some Asian countries. If IPM strategies are implemented in

It is ... in the interest of national stability and security that countries develop and implement environmentally sound agricultural development plans.

combination with the application of the International Code of Conduct on the Distribution and Use of Pesticides, and the training of farmers, a great deal will be achieved in reducing the environmental impacts of pesticides (see also Chapter 10).

Several biotechnologies have been recently developed to solve specific agricultural problems. For example, the herbicide atrazine is used to kill weeds in maize fields. Maize can tolerate atrazine. However, where maize is planted in rotation with soybeans, the latter are susceptible to residues of atrazine and their yield is affected. An atrazine-resistant soybean has been developed for growing in rotation with maize. Marked progress has been made in transferring the genes for nitrogen fixation present in certain bacteria to some crops, which would lead to dramatic improvements in biological nitrogen fixation

Environmentally sound forms of pest control

IPM in Asia

In the 1970s, the development of high-yielding strains of rice and increased use of fertilizers and pesticides allowed farmers in Indonesia to grow two rice crops each year instead of one. Unfortunately, this led to an enormous growth in the population of brown plant hoppers. Farmers were spraying up to eight times in the rice-growing season to try to reduce the damage done by this pest, and the government was providing large subsidies to help the farmers pay for expensive pesticides.

Then scientists showed that spraying had caused the problem in the first place. The sprays had wiped out all the natural predators of the brown plant hoppers, particularly spiders, and yet had had only a limited effect on the pest itself.

In response, the Indonesian government introduced an integrated pest management (IPM) system. First, it reduced the subsidies on chemical sprays, and banned farmers from using 57 insecticides on rice. It then set up a nation-wide training programme to show farmers how to conserve natural predators such as spiders. Spraying was to be considered only as a last resort.

Within three years, farmers were using 90 per cent less pesticides, with large savings in cost both for them and the government. Yields of rice were increasing, and less harm was being done to the environment.

Similar IPM programmes for rice are being introduced in Bangladesh and India.

Checking the New World screw worm fly

A lethal pest—the New World screw worm fly—recently killed more than 12 000 animals in Libya. Unchecked,the larvae of the fly could have eaten their way through 70 million head of livestock in five North African countries. The outbreak began in Libya in 1988 and the flies infected about 40 000 km^2 in Libya. Female screw worm flies lay their eggs in open animal wounds. The maggots that develop eat living tissue and can eventually kill the host animal. FAO and IFAD started an eradication programme that relied on swamping the female screw worm flies with male flies that were irradiated to make them sterile. The males mate with females whose eggs then fail to hatch, and the population eventually dies out. More than a billion sterilized Mexican flies were flown into the area and used in the control operation. No infected animals have been found since April 1991.

Sources (29, 30)

Advanced biotechnology is extensively dominated by private sector research and development, and the transfer of such technologies to developing countries will be complicated and costly.

and decrease the dependence on chemical fertilizers (25). However, advanced biotechnology is extensively dominated by private sector research and development, and the transfer of such technologies to developing countries will be complicated and costly.

ELCOR
CHEMICAL CORPORATION
TX 92151

Chapter 12

Industry

One of the most visible results of development is the enormous growth of industry. The world manufactures seven times as many goods and produces three times as many minerals today as it did in the 1970s (1). Although industrial production grew rapidly between 1950 and the early 1970s, at 7 per cent annual growth, it has since slowed to about 3 per cent per year. Industry's contribution to the gross domestic product (GDP) of low-income countries increased from 28 per cent in 1965 to 37 per cent in 1989. In middle-income countries, it grew from 34 per cent in 1965 to 36 per cent in 1989. In industrial market economies, on the other hand, the contribution of industry to GDP decreased from 42 per cent in 1965 to 35 per cent in 1989 (2, 3, 4). This can be attributed to the general downturn and stagnation in industrial output in these countries since the early 1980s.

The developing countries' share of world manufacturing output remained virtually stagnant at around 12.7 per cent during the period

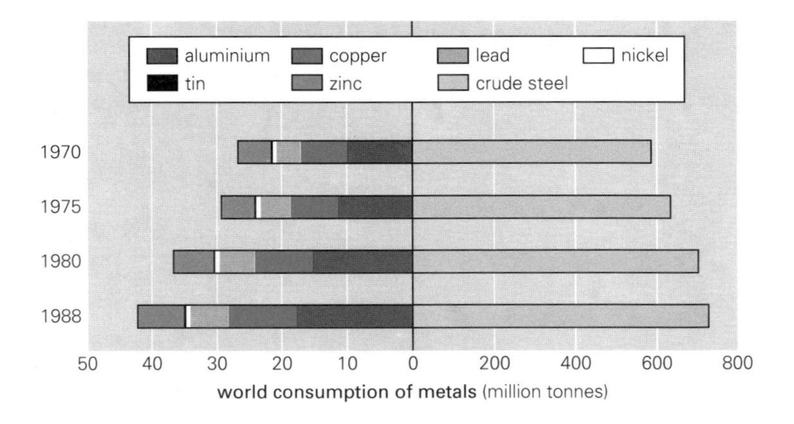

Figure 12.1
World consumption of metals, 1970–88

source (36)

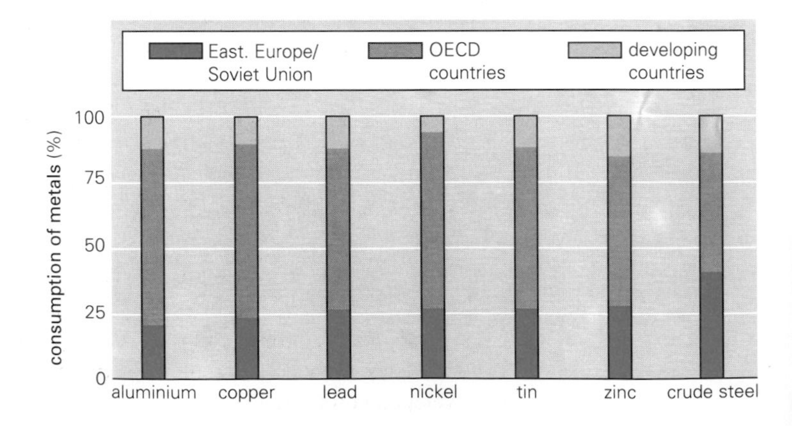

Figure 12.2
World consumption of metals by region

source (36)

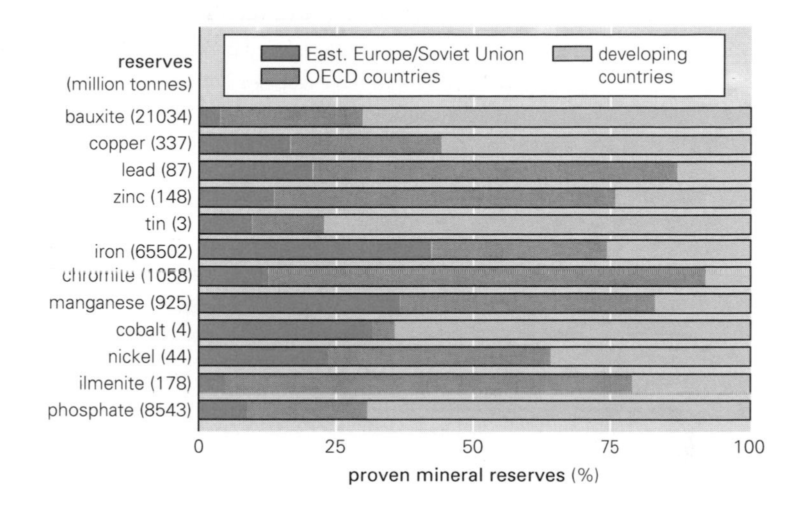

Figure 12.3
Proven reserves of minerals, by region

source (36)

1980–85, but increased slightly to about 14 per cent in 1990 (5, 6). Developing countries are beset with problems that greatly impede industrial growth. These include a rising burden of debt servicing, net capital outflows, protectionist barriers against entry into the markets of developed countries and urgent demands to meet the rising needs of their people.

The industrial sector is dynamic and rapidly evolving. The emergence of new technologies is one of the most important recent trends in industrial development. Robotics, automation, micro-electronics, information technology, new materials and biotechnology have provided the basis for and driving force behind both the development of new high technology industries and the modernization of existing production processes in traditional industries such as textiles and pulp and paper. Other important trends have been the growing substitution of one material by another, and an increase in recycling.

Industry, resources and environment

The industrial sector is an important user of natural resources and is the major contributor to the world's pollution loads. The use of metals increased over the past two decades (Figure 12.1) although regional consumption patterns vary (Figure 12.2). The developing countries have most of the world's proven reserves of important minerals such as bauxite, copper, tin, cobalt and phosphates (Figure 12.3) but they consume only about 12 per cent, exporting most of their production to developed countries.

The extraction of minerals (and their concentration and initial processing) has several negative impacts on land, water and the atmosphere. These impacts are especially large in developing countries where mining operations are generally less sophisticated than those in developed countries, and mostly lack environmental protection measures. For example, bauxite processing in Jamaica produces massive quantities of 'red mud' which has contaminated groundwater resources (7). The mining of tin, copper, phosphates and iron ores has also created water and air pollution problems in some African and Asian countries. Years of mineral extraction in some countries—for example the United States—without due consideration to environmental impacts have created large areas of wasteland and massive amounts of accumulated hazardous wastes (Chapter 10).

One important feature of the past two decades has been the growing substitution of one mineral resource by another or by non-mineral material, basically to reduce costs and to save weight and, consequently, energy. The automobile industry, for example, achieved a weight saving of about 25 per cent over the past decade through the use of plastics, ceramics, aluminium and ultra-strong sheet metal. The aircraft industry uses carbon fibres and ultra-light alloys of aluminium and lithium to achieve the same goal (8). Another reason for material substitution is the superior technical properties offered by the new material. In communications, for example, glass fibre is superior to conventional copper cable in nearly every application. A communications satellite weighing 250 kg performs better than a trans-oceanic telephone cable weighing 150 000 tonnes. Material substitutions have lead to a decrease in the consumption of, for example, crude steel and an increase in the use of, for example, aluminium (Figure 12.4).

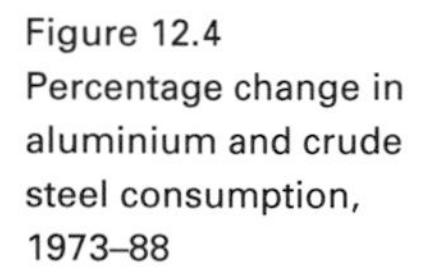
Figure 12.4 Percentage change in aluminium and crude steel consumption, 1973–88

source (36)

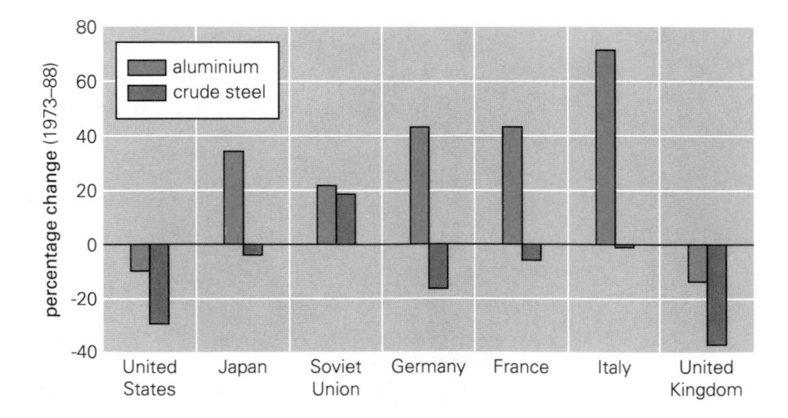

Material substitution has its own environmental impacts, substituting the environmental impacts associated with the extraction and processing of one mineral for those of the new material. In some cases, the latter are more severe. For example, the processing of semi-conductors, optical fibres, new classes of ceramics, and composites requires large quantities of toxic chemical compounds, which create significantly greater health and safety problems for workers and the public, especially when accidents occur. Another important issue is that most such new materials cannot be easily decomposed and the disposal of their waste products can create problems never previously encountered (9).

Worldwide, industry consumed about 540 km^3 of water in 1970 (about 21 per cent of total global freshwater withdrawal), and about 973 km^3 in 1990 (24 per cent of total withdrawal)—an increase of about 80 per cent in two decades (Chapter 5). This amount is expected to reach 1280 km^3 in 2000, constituting about 25 per cent of total freshwater withdrawal worldwide. The modest amounts of water used in the industrial sector, compared to water withdrawn for agriculture, is due to the fact that many industries reuse water several times before it is finally discharged as industrial wastewater. For example, in the United States, each cubic metre of water is used on average about nine times before being finally discharged as wastewater. This rate of water reuse is expected to reach 17 times in 2000 (10). Such water reuse varies between industries and countries, and depends on the cost of water and its availability, and on the cost of recycling. In some countries, for example India, Japan and Germany, treated domestic wastewater (sewage) is used in some industries for cooling or as process water.

Industry consumes more energy than any other end-use sector, accounting for 37 per cent of worldwide commercial energy consumption in 1990. Differences exist, however, between countries. In the OECD countries, the average percentage is 33; in Eastern Europe it is 60 per cent (11, 12); and in the developing countries, it varies from a low of 11 per cent in Uganda to a high of 69 per cent in China (13). An important development in the past two decades has been the marked decline of industrial energy intensity (the ratio of industrial energy use to value added) in most OECD countries

Glass fibre is superior to conventional copper cable in nearly every application. A communications satellite weighing 250 kg performs better than a trans-oceanic telephone cable weighing 150 000 tonnes.

(14, 15, 16). Energy efficiency improvements appear to have been the major cause of decreased industrial energy intensity, although structural changes within the industrial sector also played a very important role (Figure 12.5). In contrast, industrial energy intensity in Eastern Europe has remained either constant or declined only slightly. Measures to increase industrial energy efficiency in developing countries have been of very limited success. Industry there often consumes two to five times as much fuel for a given process, due to decades-old industrial equipment (13, 17). In some countries, industrial energy subsidies and the requirements to produce fixed quotas of goods at fixed prices have deterred efforts to improve energy efficiency. However, a study in Thailand (18) has shown that better industrial 'house keeping' alone could lead to a 12 per cent improvement in energy efficiency, and process improvement could add an additional 16 per cent. Another study in Egypt (19) showed that improved industrial house keeping could conserve 20 per cent of industrial energy use.

Impacts of industry on the atmosphere

Industry emits many air contaminants. The quantities and types of compounds emitted depend mainly on the type of industry, the raw material, fuel and technology used, and the environmental protection measures in place. Factors such as the size of the industrial installation, age of machinery, standard of maintenance and management are also important. In addition to the common air pollutants such as sulphur and nitrogen oxides, carbon dioxide, carbon monoxide, hydrocarbons and particulate matter, industry emits hundreds of trace contaminants, some of which are potentially toxic (Chapter 10).

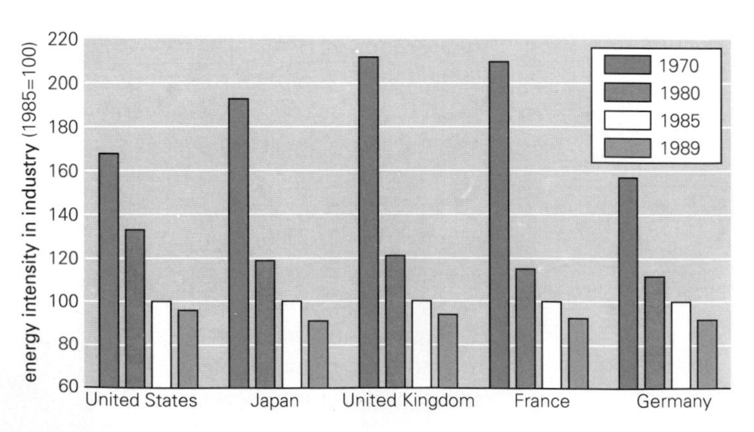

Figure 12.5 Energy intensity in industry

based on data from (37)

Estimates of major emissions into the atmosphere due to industry

	million tonnes per year	*per cent of anthropogenic emissions*
carbon dioxide	3500	50
methane	84	24
nitrous oxide	0.2	13
ammonia	7	20
sulphur oxides	89	90
nitric oxides	30	44
particulate matter	23	40
hydrocarbons	26	50
chlorofluorocarbons/ halons	1.2	100

Includes utilities (power and steam-generating stations). Chlorofluorocarbons and halons represent 1986 level (Chapter 2).

source (38)

The size of the emissions vary greatly. In 1989, industry in the OECD countries was responsible for 25 per cent of NO_x emissions, 40-45 per cent of SO_x and 50 per cent of total greenhouse gases (20). More detailed statistics from the United Kingdom (21) show that industry in 1988 was responsible for 91 per cent of SO_2 emissions, 47 per cent of NO_x, 60 per cent of CO_2 and 3 per cent of carbon monoxide (utilities such as power stations and steam generation are normally included in the industry sector). In Hungary, industry is responsible for 94 per cent of SO_2 emissions and for all emissions of chlorine and fluorine (22). The box on this page gives the calculated estimates of contributions of industry to global anthropogenic air emissions. The impacts of these emissions on the environment are discussed in Part I.

Impacts of industry on water

The use of water in industrial processes produces billions of cubic metres of industrial wastewater daily. These wastewaters vary in composition from those similar to municipal sewage (but often more concentrated) to those which are more toxic and contain a great variety of heavy metals and synthetic organic compounds. In 1989 industrial wastewater from OECD countries contributed about 60 per cent of the biological oxygen demand load of surface waters receiving discharges, and about 90 per cent of the toxic substances load (20). Industrial wastewater discharged without adequate treatment into surface waters has created serious environmental problems that have affected aquatic life, in particular when accidental releases occurred

A study in Thailand has shown that better industrial 'house keeping' alone could lead to a 12 per cent improvement in energy efficiency, and process improvement could add an additional 16 per cent.

(see Chapters 4, 5 and 10). In several countries, wastewater from some industries has been discharged to public sewers under the pretext that such wastewater contains mainly biodegradable material that can be treated together with sewage in treatment plants. However, the uncontrolled discharge of industrial wastewater, especially that containing toxic compounds, into municipal sewers could stress and completely destroy the microbial-based systems used to treat domestic wastes. Then neither the industrial nor the municipal wastewater is effectively treated. In addition, the sludge produced from treatment plants would contain high concentrations of toxic contaminants that would be difficult to manage.

Solid wastes

Worldwide, industry generated about 2100 million tonnes of solid wastes and 338 million tonnes of hazardous wastes in 1989. Of these, 68 per cent of the former and 90 per cent of the latter were generated in OECD countries (20). Most solid wastes are generated by the metallurgical, building and chemical industries, especially during the extraction and processing of raw materials. Although some industrial solid wastes are considered 'inert' and can be treated and disposed of like urban solid wastes, others (especially hazardous wastes) require special management techniques. In Italy, for example, of about 35 million tonnes of industrial solid wastes generated every year, 40 per cent are recycled, 46 per cent are treated as inert waste in ways similar to urban refuse, and the remaining 14 per cent require special handling and treatment (23). In Spain, industry generates about 10 million tonnes of solid wastes every year, about 9 million of which are considered inert; the remaining 1 million tonnes are considered hazardous. The management of industrial solid wastes, especially the hazardous ones, remains a problem in many countries, although there are several opportunities to use many of the wastes in beneficial ways. For example, collected fly and bottom ash from power plants has been used for the manufacture of bricks and for road building in some East European countries (see also Chapter 10, especially on transboundary movement of hazardous waste).

Emerging issues

The location of industrial installations became an issue of concern following several serious accidents in the past two decades (Chapter 9). Although some countries have encouraged the dispersion of industrial installations, others have preferred to concentrate industries in 'industrial estates'. Interest in controlling pollution has been a contributing factor, but rarely the decisive one, in the choice of approach (24). Industrial dispersion policies are most often viewed as a means of distributing resources, markets and employment, and of diverting population growth from overcrowded urban centres. On the other hand, the establishment of industrial estates, for example, in Brazil, Colombia, Mexico, the Republic of Korea and Thailand, was mainly based on economies of scale in the construction of infrastructure. However, the economic benefits have been partially offset by environmental and health hazards created by the estate or industrial district itself. For example, in the industrial district of Cubatão, Brazil, where 23 major industrial plants and numerous small operations are concentrated, serious health problems, including an elevated neonatal mortality rate, birth deformities, and a high prevalence of respiratory disorders, have been associated with high levels of water and air pollution (25, 26). Recent environmental management measures undertaken in the district have now helped reduce industrial emissions. Industries that were constructed five or ten years ago beyond city limits, in many developing countries, have now become part of, and have contributed to, urban sprawl. Given scarce resources and limited transport, new migrants and the low-wage labour force attracted to urban areas have no alternative but to settle dangerously near the plants. The high death toll of accidents such as those at San Juanico in Mexico and Bhopal in India has been mainly attributed to high population densities in squatter settlements around the plants (Chapter 9).

Many analysts in the 1970s and 1980s predicted that industries would relocate in developing countries following strict environmental regulations in the industrialized countries (27, 28). There is little evidence that this has taken place on the scale predicted, although some developing countries have promoted investment by loosening environmental controls (24, 29, 30). In general, the decision to relocate an industry has been based on economic factors, but in some cases these economic factors may—in turn—have been based on environmental ones. Industries that prove unprofitable in industrialized countries may find relocation to developing nations attractive,

Industry generated about 2100 million tonnes of solid wastes and 338 million tonnes of hazardous wastes in 1989. Of these, 68 per cent of the former and 90 per cent of the latter were generated in OECD countries.

because they are able to avoid costly health and safety measures in their new locations. The loosening of safety standards has been responsible for serious accidents. Numerous countries in South-east Asia, Latin America and the Caribbean have developed 'free zones' in which foreign-owned firms can process chemicals and raw materials, and assemble and manufacture goods, all the final products being exported. Because environmental protection measures are not stringent in these areas, they may have serious impacts on the health of workers, the surrounding population, and the environment (31).

Focus has, hitherto, been on the environmental impacts of large industries. In many developing countries, small-scale industries have grown substantially in type and in number, and account in some countries for more than 60 per cent of the labour force, many of whom are women and children. Textiles, garment and footwear manufacture, auto-repair, gem polishing, foundry work, scrap processing and rubber curing are among the hundreds of expanding small industries in urban and rural areas. In China, for example, about 18 per cent of the national industrial output in the 1980s came from small-scale industries involving nearly 7 million workers. During the next 20 years, planners estimate that 150 million people will be employed in small-scale industries in China (32). Although small-scale industries contribute to national economies and improve the living conditions of many people, they also carry increased health and environmental risks. Studies in Thailand and Brazil, for example, have shown that the prevalence of occupational diseases is higher in small-scale than in medium-scale or large industries (33, 34). Women and children are particularly vulnerable (31). Nearly all small-scale industries dispose of their liquid wastes without treatment into public sewers or nearby surface waters. Solid wastes are often dumped with domestic refuse. The cumulative effects of these wastes could in some cases surpass those from large industries.

Since the late 1970s the world has experienced a technological revolution propelled by extraordinary scientific progress and rapidly advancing technology. Computers, telecommunications, biotechnology, lasers and new materials have brought the global economy to the threshold of a new industrial age. Our knowledge of the environmental impacts of these new technologies is still, however, in its infancy. For example, although industrial applications of genetic engineering will be subject to strict safety measures to ensure that genetically engineered organisms are contained, we do not know what may happen if such organisms are accidentally released into the environment. The deliberate release into the environment of organisms for agricultural or environmental purposes may cause health hazards and/or damage to particular ecosystems that cannot be controlled. If 'prevention is better than cure', then information on new technologies should be widely disseminated to enable the risks to society and the environment to be assessed and hence to help identify the gaps in knowledge which call for further research by the scientific community. Once the risks have been assessed, suitable measures could be formulated to deal with them and prevent or minimize possible hazards before they occur.

Responses

The traditional model of industrial activity—in which individual manufacturers take in raw materials and generate products to be sold plus waste to be disposed of—is now being gradually transformed into a more integrated model—'an industrial ecosystem'. In such a system the consumption of energy and materials is optimized, waste generation is minimized and the effluents of one process serve as the raw material for another (35). This 'greening' of industry is demonstrated by achievements in several OECD countries in increasing the efficiency of energy and water use, increased recycling of waste, and the development of cleaner technologies (Chapter 10). The cooperation of the chemical industry has been an important driving force behind the steps taken to phase out chlorofluorocarbons and other compounds that deplete the ozone layer (Chapter 2). By this and by increasing the efficiency of energy use, industry is helping reduce emissions of greenhouse gases. There have been also several successes in reducing emissions of sulphur oxides (Chapter 1) and in treating industrial wastewater, with consequent improvements in the quality of such rivers as the Rhine and the Thames. Growing international concern over transfrontier movements and dumping of hazardous wastes, especially in the developing countries, led to the adoption in 1989 of the Basel Convention on the Control of Transboundary Movements of Hazardous Wastes and Their Disposal (Chapter 10).

In an industrial ecosystem, the consumption of energy and materials is optimized, waste generation is minimized and the effluents of one process serve as the raw material for another.

Chapter 13

Energy production and use

The demand for energy has increased dramatically this century (Figure 13.1). During the first half of the century, global energy consumption grew by about 2.2 per cent a year; between 1950 and 1970, this increased to 5.2 per cent; and between 1970 and 1990 it slowed to 2.3 per cent. The slower increase over the past 20 years resulted from, among other things, increases in oil prices in the early and late 1970s and the introduction of measures to increase the efficiency of energy use and to curb the rising demand for energy in the developed countries.

There have also been major changes in the energy mix used. A century ago, non-commercial sources of energy (such as fuelwood, agricultural residues and dung) constituted about 52 per cent of total energy used. That share dropped significantly as fossil fuels became the predominant source of energy. In 1930, non-commercial fuels comprised 25 per cent of total energy used, in 1950, 21 per cent, and in 1970, 12 per cent. This figure has since remained almost

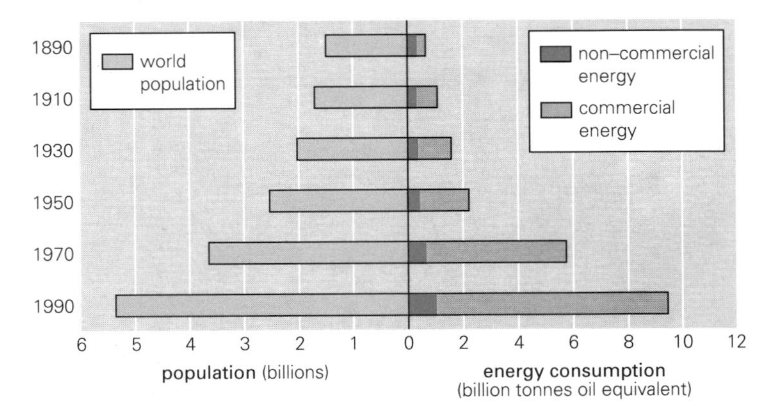

Figure 13.1
Energy consumption in the world

based on data from (29)

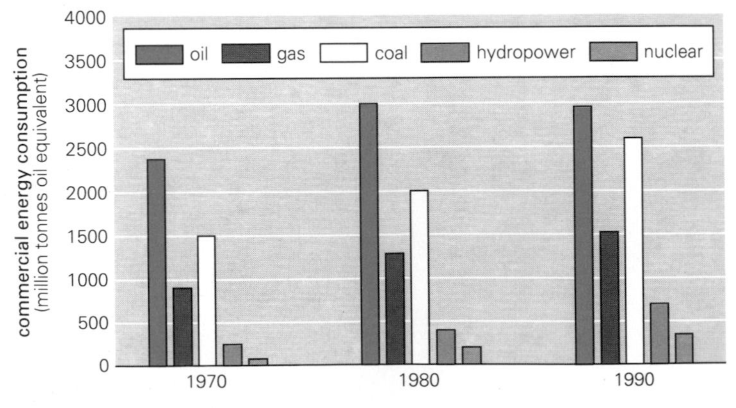

Figure 13.2
World consumption of commercial energy, by source

based on data from (30)

unchanged, although more than 2 billion people in the developing countries still depend on non-commercial fuels, especially fuelwood, for a substantial part of their energy needs (Chapter 7).

Another important change has been the decline of the share of coal. In the 1920s, coal accounted for about 80 per cent of the world's total commercial energy consumption; in later decades it was displaced mainly by oil. In 1970 coal accounted for 29 per cent of total energy consumption, and its share increased slightly to 32 per cent in 1990 (Figure 13.2). Oil has been the main source of energy over the past two decades, although its share declined from 46 per cent in 1970 to 36 per cent in 1990.

The consumption of commercial energy is heavily concentrated in OECD countries, Eastern Europe and the Soviet Union (Figure 13.3). In 1990, these countries, with about 22 per cent of the world population, consumed about 82 per cent of the world's commercial energy, whereas the developing countries, with 78 per cent of the world population, consumed only about 18 per cent. On average, a person living in the high-income countries consumes 15 times more energy than one living in the low-income countries (Figure 13.4), and about 4 times as much as one living in the middle-income countries (1). Wide disparities exist, however, among different groups of people in the same country.

Events in the past two decades brought home a general realization that the era of cheap energy was over and that all economies would have to adapt to high energy prices. Furthermore, the fact that supplies of fossil fuels are finite became more evident than ever. This has brought into focus the importance of establishing energy mixes to meet demand, with more reliance on indigenous resources, and the importance of increasing the efficiency of energy use. By the end of the 1980s, it had become clear that current trends in energy consumption—especially of fossil fuels—could lead to increased degradation of the global environment (from, for example, acid rain, urban air pollution and climate change),

Figures 13.3 and 13.4 World consumption of commercial energy, by region (below) and per capita energy consumption (bottom)

based on data from (30) and (1) respectively

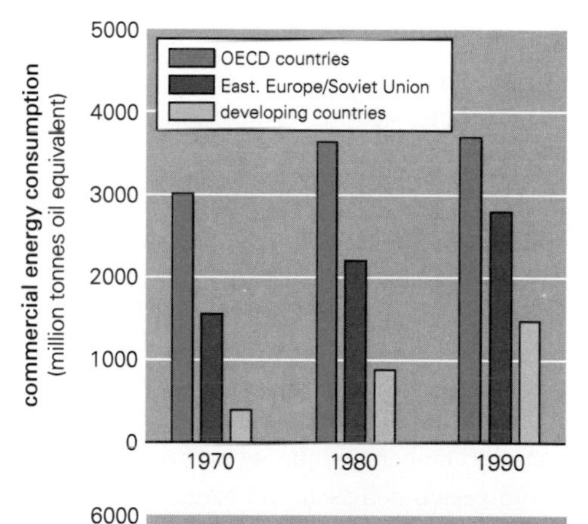

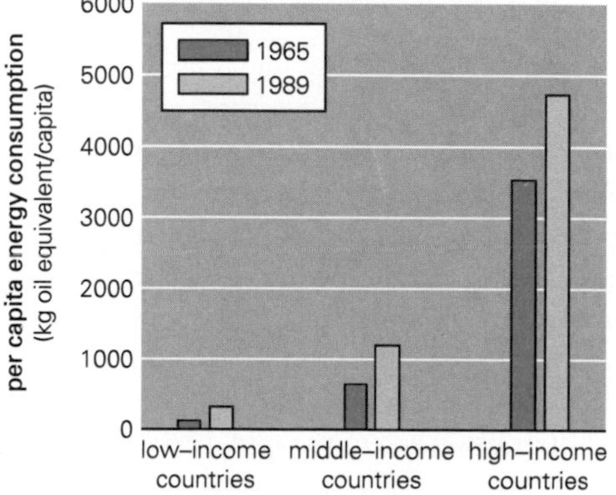

undermining future development and well-being across the planet.

Many projections have been made of future world energy demand (2, 3, 4) but the assumptions on which they are based include many uncertainties. There is general agreement, however, that world energy demand will continue to rise; the growth rate will be highest in developing countries (about 4.5 per cent per year), followed by Eastern Europe and the Soviet Union (3 per cent per year) and lowest in OECD countries (about 1.3 per cent per year). Although the OECD's share of total world energy consumption is likely to fall from 46 per cent at present to about 43 per cent by 2000, it will continue to be the region with the highest consumption of energy, especially of fossil fuels (5).

At the end of 1989, the estimated proven recoverable reserves of oil in the world were 139 billion tonnes of oil—77 per cent in OPEC countries, 12 per cent in other developing nations, 6 per cent in Eastern Europe and the Soviet Union, and 5 per cent in developed market economies (6). Coal resources are about 534 billion tonnes of oil equivalent and natural gas resources 104 billion tonnes of oil equivalent. At the world's 1990 level of consumption, oil reserves would last for about 46 years, coal for about 205 years and natural gas for about 67 years.

The uneven distribution of the world's fossil fuels has created a huge worldwide trade in energy commodities: some 44 per cent of oil, 14 per cent of gas and 11 per cent of coal are traded internationally (7). Extensive distribution systems exist to serve this trade and ensure that resources reach the consumer. Natural gas is transported over land through some one million kilometres of pipelines, and oil through 400 000 kilometres of pipes, excluding local distribution systems. About 2600 tankers ply the world's oceans, carrying crude oil; another 65 vessels deliver liquid natural gas around the world.

Worldwide, a substantial amount of fossil fuel is used to generate electricity. Of about 11 000 TWh (1 terawatt-hour is 1 billion kWh) generated in 1989, some 62 per cent were from thermal power stations (fired by fossil fuels), 20 per cent from hydropower, 17 per cent from nuclear power stations, and less than 1 per cent from geothermal resources. The share of nuclear power grew from 1.6 per cent in 1970 to 16.8 per cent in 1989 (8). As of 31 December 1990, there were 423 nuclear power plants in the world, with a total of 325 873 megawatt-electric (MWe) of installed nuclear power generating capacity (9). Projections made in the mid-1970s that nuclear power would contribute 2600 gigawatt-electric (GWe) by the year 2000 were revised and scaled down to 1075 GWe in the early 1980s, and then to 444 GWe according to an IAEA 1987 projection (10). In the OECD countries, nuclear power accounted for an average of 22 per cent of electricity generation in 1987, fossil fuels for 60 per cent, and hydropower and geothermal energy for the remaining 18 per cent. These proportions are projected to remain unchanged until the year 2005 (11). However, the mix of fossil fuels used for electricity generation will change; the share of oil will decrease and will be displaced by coal.

Energy production, transformation, transport and use have important impacts on the environment. These impacts depend on the source of the energy, the technologies used in its production, and the sector involved: agriculture, industry, transport, and domestic and commercial. The environmental impacts of different energy systems are normally assessed for the entire fuel cycle—from extraction of raw material, through transportation, processing, storage and use of the fuel, to the management of wastes generated in all steps of the cycle. Such environmental impacts have been the subject of extensive studies by UNEP and other organizations since the mid-1970s-see, for example, (12-21) for details on the environmental impacts of different sources of energy. The following focuses on issues that are of major concern.

Impacts of energy production and use on atmosphere

The combustion of fossil fuels and biomass generates air pollutants which vary in type and quantity according to the fuel used. Fossil fuel combustion accounts for the release of about 90 per cent of global anthropogenic sulphur oxides, 85 per cent of nitrogen oxides, 30–50 per cent of carbon monoxide, 40 per cent of particulate matter, 55 per cent of volatile organic compounds, 15–40 per cent of methane and 55–80 per cent of carbon dioxide (21). The shares of emissions in the different sectors vary widely from one country to another and depend on the amount and composition of fuel used and on the emission abatement technologies in place. Coal combustion emits more sulphur oxides, nitrogen oxides and carbon dioxide per unit of energy

At the world's 1990 level of consumption, oil reserves would last for about 46 years, coal for about 205 years and natural gas for about 67 years.

than oil, natural gas and biomass. On the other hand, biomass burning emits more carbon monoxide than coal, oil and natural gas. The trends and impacts of emissions of sulphur oxides, nitrogen oxides, carbon monoxide and particulate matter from stationery and mobile sources are discussed in Chapter 1. Of particular concern are the impacts of different emissions on urban air quality, and their role in acidic deposition and climate change.

The amount of carbon dioxide emitted as a result of energy use in 1988 was about 6.3 billion tonnes of C (oil combustion contributed about 2.4 billion tonnes C, natural gas 1.1, coal 2.4, and non-commercial fuels about 0.5). If current trends of energy use and efficiency prevail, carbon dioxide emissions will reach about 9.1 billion tonnes of C in 2005 (5) and may double by 2010 (7). However, if energy can be used more efficiently, the amount of carbon dioxide emitted in 2010 may be only 50 per cent higher than in 1988. If energy efficiencies can be radically improved, the increase will be only 15 per cent.

Impacts of energy production and use on water

Water pollution can result from several energy-related activities. Acid mine drainage has polluted surface water streams in the United States and several other countries, and has reduced or eliminated aquatic life in many of them. Marine pollution has occurred from normal discharges of ships and offshore oil platforms, and from accidental oil spills (Chapter 9). Oil refineries discharge liquid effluents containing oil, grease, phenols, ammonia and other toxic compounds. Power plants use water for cooling and the discharged water is usually about 7 °C warmer than the receiving water bodies. Such thermal pollution has been claimed to affect aquatic life, but in some countries these thermal waters have been used for aquaculture, irrigation and other purposes (12).

Impacts of energy production and use on land

Coal mining, especially strip mining, disturbs large areas of land. Although strip-mined areas have been been successfully reclaimed in some countries, for example Germany, there is concern that future increases in coal utilization could disturb more land and affect human settlements near mining areas (12, 22). All other energy-related activities require land which may not be readily available, or which might be better used for other purposes. The building of a dam may inundate forest areas, with detrimental effects on wildlife; the construction of windmill parks or of solar power stations requires extensive areas and may compete with other land uses; and energy plantations may compete with land use for food production. Land is also required for the management of massive amounts of solid wastes generated in some fuel cycles (especially coal and nuclear). Mining and processing of coal and uranium leave considerable amounts of solid wastes that must be properly disposed of. A major and growing source of solid wastes has developed along with air pollution control measures at power plants fired by fossil fuels. Sludge from flue gas desulphurization and ash collected by electrostatic precipitators, in addition to bottom ash, add to the problem of solid waste management which requires increasing land areas.

Nuclear power and the environment

Concern about nuclear power development has focussed on a number of issues, the most important of which are: the effects of radiation on humans, the safety of nuclear installations, the environmental impacts associated with radioactive waste management (including the decommissioning of nuclear installations), and the possibilities of diversion of nuclear material for non-peaceful uses (12, 15). At each stage of the nuclear fuel cycle—from the mining and milling of uranium ores to fuel fabrication, power plant operation, eventual reprocessing of irradiated fuel and the disposal of nuclear wastes—radioactive materials are released into the environment. The radionuclides released decay at different rates; most are of only local importance because they decay rapidly; some live long enough to spread right around the world; and some remain in the environment virtually for ever. Different radionuclides also behave differently in the environment; some spread quickly, others move very little. In all, the operation of the nuclear fuel cycle contributes about 0.04 per cent of all radiation to which humans are exposed; natural sources, in comparison, account for 83 per cent and medical sources for about 17 per cent (23).

People living near nuclear installations do, of course, receive much higher doses than the average. Even so, typical doses around nuclear reactors at present

The amount of carbon dioxide emitted as a result of energy use in 1988 was about 6.3 billion tonnes of carbon ... If current trends ... prevail, carbon dioxide emissions will reach about 9.1 billion tonnes of carbon in 2005 and may double by 2010.

constitute a fraction of one per cent of doses from natural sources. All these figures assume that the nuclear plants operate normally since very much larger quantities of nuclear material may be released in accidents (Chapter 9). Although a vast amount of information has accumulated about the acute effects of radiation, many uncertainties remain about the effects of low-level radiation (12, 15, 23, 24). The difficulty of proving cause and effect is also a problem in studying links between human genetic effects and irradiation. A recent study (25), however, revealed that the incidence of leukaemia was higher in children born near the Sellafield nuclear plant in the United Kingdom and in the children of fathers employed at the plant, particularly those with high radiation dose recordings before their children's conception.

Figure 13.5
Trends in energy intensities by region (top) and for some selected countries (bottom)

based on data from (6, 27, 31)

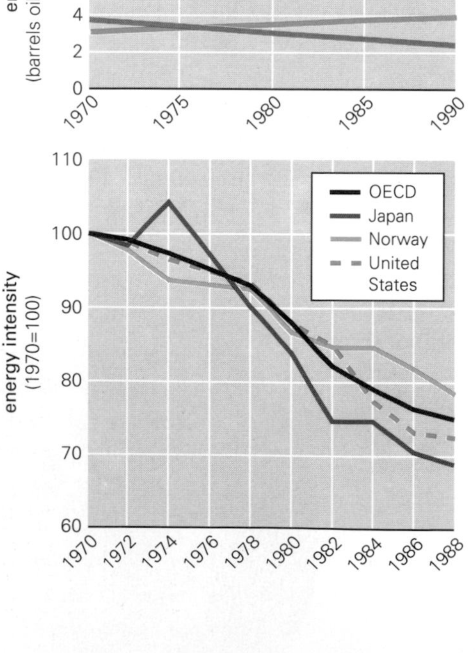

Radioactive wastes are generated at all steps of the nuclear fuel cycle. The bulk of the wastes occur at the beginning of the cycle which includes mining and milling, while the more radioactive wastes occur near the end of the cycle which includes reactor operation and fuel reprocessing. The latter wastes are generally divided into low-level wastes (LLW), intermediate-level wastes (ILW) and high level wastes (HLW) which include wastes from reprocessing plants and/or spent fuel from nuclear reactors. Worldwide, the volume of LLW generated in 1990 was about 370 000 m^3, that of ILW about 27 000 m^3 and that of HLW and spent fuel about 21 000 m^3 (26). By the year 2000, the cumulative amount of LLW from nuclear power reactors could reach some 7 million m^3, and that of HLW about 1 million m^3 (12). LLW are normally disposed of in surface, shallow or underground installations, which should be controlled for about 300 years (27). ILW are generally conditioned in cement, bitumen or resin, and buried underground in shallow repositories. No HLW have yet been disposed of. National authorities are storing them; and some have been researching ways of solidifying them and disposing of them in stable geological formations on land and on, and under, the seabed (23, 27).

Decommissioning of nuclear installations—the process of dismantling and disposing of old nuclear plants—is technically feasible. However, the issues involved in decommissioning are complex: the technical, economic, radiological, environmental and organizational can conflict, and these conflicts cannot be resolved until waste disposal routes are defined. The IAEA reported in 1990 that 143 nuclear facilities (116 research reactors, 16 power plants and the rest other facilities) in 17 countries were at some stage of decommissioning (no large nuclear power plant has yet been decommissioned). Moreover, 64 nuclear reactors and 256 research reactors could become in need of decommissioning by the year 2000 (28). The costs of decommissioning are high, estimated (perhaps conservatively) at about US$480 million for a 1000 MWe nuclear plant.

Responses

In many countries, especially OECD countries, the environmental impacts of production and use of energy have been reduced over the past two decades, as a result of more efficient use of energy, changes in the energy mix used, and control of emissions. The least progress has been in developing countries where energy conservation has been weak in all sectors and financial problems have constrained investment in emission control. Overall, energy intensity (energy use per unit of gross domestic product, GDP) in the developed market economies declined by 29 per cent between 1970 and 1990; in Eastern Europe and the Soviet Union, it declined by 20 per cent, although energy intensity is about three times higher than it is in developed market economies. In contrast, the energy intensity in developing countries increased by 30 per cent between 1970 and 1990 (Figure 13.5). This has been mainly attributed to the very limited success of efforts to increase the efficiency of energy use and the fast replacement of non-commercial fuels by commercial sources such as oil products and electricity. It is expected that energy intensity will continue to decline in OECD countries by about 1.3 per cent per year until 2000 (5), but will not change in East Europe and the Soviet Union, and may continue to rise in developing countries (5, 6).

The change of the energy mix, especially in OECD countries, and the introduction of

The IAEA reported in 1990 that 143 nuclear facilities in 17 countries were at some stage of decommissioning (no large nuclear power plant has yet been decommissioned). Moreover, 64 nuclear reactors and 256 research reactors could become in need of decommissioning by the year 2000.

emission control measures have led to a marked decrease in emissions of sulphur oxides and carbon monoxide (Chapter 1) in the past two decades. Although there is a trend in OECD countries towards strict regulation of new large combustion facilities, regulation of existing power plants is less consistent because retrofitting existing plants is in some cases not cost effective. Advanced generating technologies which offer a number of advantages over conventional technologies (lower emissions of NO_x and SO_x, and higher thermal efficiencies) are being developed in countries (such as Germany, the United Kingdom and the United States) that are planning to increase the utilization of coal for electricity generation. The development of combined heat and power (CHP), the promotion of industrial autogeneration, the use of industrial waste heat and the development of CHP/district heating systems have also contributed to more efficient use of energy and, consequently, the reduction of emissions.

ปอ.11
ISUZU
ISUZU

Chapter 14

Transport

Transport is an essential component of social and economic development. Today, more people travel over greater distances and more fuels, raw materials and products are transported around the world than ever before. Transport systems and modes vary geographically and change continuously over time. In many developing countries, draught animals remain the principal means of conveying goods over short distances, while personal travel is predominantly on foot, particularly in rural areas. In semi-urban and urban areas, tri-shaws, cycle rickshaws, pedal carts and other traditional forms of transport account for a large share of road transport use in developing countries. In some countries, bicycles are important: in 1989 China and India together had an estimated 600 million bicycles (1); and cycling has for long been popular in The Netherlands, Denmark and some other European countries.

Road transport is the most popular mode of transport for both passengers and freight in the developed nations, and is becoming increasingly important in developing countries. The number of motor vehicles in the world more than doubled in the past 20 years (Figure 14.1) and is expected to double again in the next 20 or 30 years. Vehicle production and ownership are still overwhelmingly concentrated in developed countries. The OECD countries account for 88 per cent of car production and 81 per cent of the global fleet (2, 3). Car ownership in the developing countries has risen sharply, averaging an annual growth rate of 10 per cent per annum between 1970 and 1990, and is expected to grow further as vehicle ownership in developed countries stabilizes. The level of motorization will continue, however, to be much higher in the developed than in the developing countries. The number of cars per 1000 inhabitants in the United States is about 550, in Western Europe 200–400, in Africa 9, in India 2, and in China 0.4.

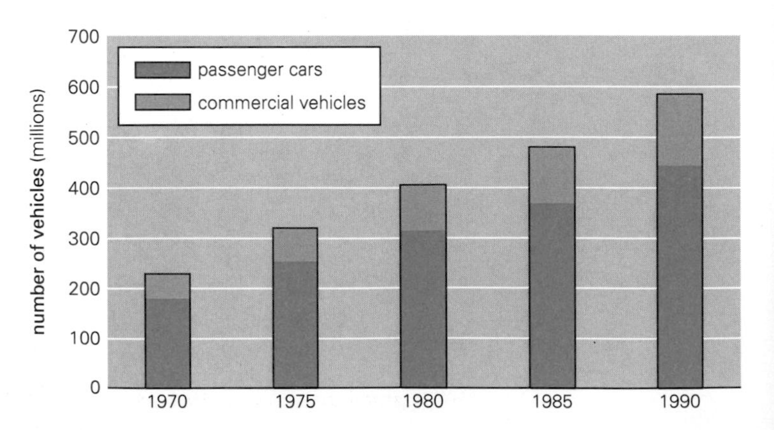

Figure 14.1
Motor vehicles in use in the world

based on data from (4, 5, 23)

Other modes of transport also grew since 1970. Civil aviation flew about 7 billion kilometres in 1970, achieving 382 billion passenger-km. These figures rose to 12 billion kilometres and 1368 billion passenger-km in 1987 (4, 5). Railway freight increased, from 5019 billion net tonne-km in 1970 to 7285 in 1987. Sea shipping rose from 2605 million tonnes in 1970 to 3675 million tonnes in 1980, but dropped to 3361 million tonnes in 1987 as a result of the decrease in oil transport, which accounts for about 55 per cent of all goods transported at sea.

Transport, resources and environment

The transport sector consumes vast amounts of resources—land for roads, railways, harbours and airports, minerals and metals for the construction of vehicles and of transport infrastructure, and substantial amounts of energy. Motorways in the OECD countries increased from 73 000 km in 1970 to 132 000 km in 1988—an increase of 81 per cent (6). In the developing countries, the construction of new motorways (and roads in general) has been constrained by economic difficulties, and over the past two decades the conditions of existing roads have deteriorated in many developing countries due to inadequate maintenance. In some countries, there has been land use conflict between the transport sector (for construction of motorways, railways, harbours and airports) and other sectors, including food production.

Worldwide, the transport sector accounts for about 30 per cent of total commercial energy consumption, of which road transport alone consumes 82 per cent, almost all of which are oil-derived products (7, 8). However, wide differences exist between regions and countries. For example, in Eastern Europe and the Soviet Union, the transport sector accounts for about 13 per cent of total energy consumption (9); in Kenya it accounts for about 45 per cent (10).

Since the early 1970s, several alternatives to oil have been studied as vehicle fuels. Attention currently centres on alcohol fuels (ethanol and methanol), natural gas and, to a lesser degree, electricity. Alcohol fuels can be derived from biomass; methanol can also be produced from natural gas and coal. Brazil's

The number of motor vehicles in the world more than doubled in the past 20 years and is expected to double again in the next 20 or 30 years.

ethanol programme (derived from sugar-cane), launched in 1975, provided roughly half the country's vehicle fuel in 1986. Almost one-third of Brazilian cars are now capable of running on pure ethanol; others are running on an 80/20 petrol-ethanol blend (2, 11). Using natural gas directly as a vehicle fuel, either in compressed (CNG) or in liquefied (LPG) form, is becoming popular in some countries. There are now more than 300 000 CNG vehicles on the road in Italy. Japan and Italy meet almost 4 per cent of their national transportation fuel demand with LPG. Other countries such as Argentina, Australia, Indonesia, New Zealand, Pakistan and Thailand are beginning to use natural gas as a transportation fuel.

Impact of transport on the atmosphere

Cars, trucks and buses play a prominent role in generating virtually all the major air pollutants, especially in cities. Petrol-burning vehicles emit carbon dioxide, carbon monoxide, hydrocarbons, oxides of nitrogen, particulates and trace compounds. In confined places and congested streets, carbon monoxide concentrations can rise to levels hazardous to health, especially to people with heart or lung weakness. Oxides of nitrogen and hydrocarbons interact in the presence of sunlight to produce an oxidant smog which irritates the eyes and lungs, and damages sensitive plants. In countries where leaded petrol is used, almost all lead in air emissions in cities is from vehicle exhausts. Studies carried out near highways have shown elevated concentrations of trace metals such as cadmium, lead, copper, zinc, nickel and chromium on vegetation and soil (11, 12). Although diesel-powered vehicles emit comparable or less carbon monoxide and hydrocarbons than petrol-powered cars, they emit 30–50 times more particulate matter (13, 14). Some 80–90 per cent of such particulates are less than one micrometre in diameter and, hence, are easily transported by air currents, and readily settle in the lower respiratory tract when inhaled. These particulates contain hundreds of organic compounds, several of them carcinogenic. Aircraft and railway locomotives together emit a smaller volume of air pollutants than do road vehicles. However, it has been estimated that the world's fleet of civilian aircraft generates about 2.8 million tonnes of nitrogen oxides per annum, which could enhance the formation of tropospheric ozone (15).

Worldwide, the transport sector generates about 60 per cent of anthropogenic carbon monoxide emissions, 42 per cent of nitrogen oxides, 40 per cent of hydrocarbons, 13 per cent of particulates and 3 per cent of sulphur oxides (see box). The transport sector is also a

Estimates of major emissions into the atmosphere from transport

	million tonnes per year	*per cent of anthropogenic emissions*
carbon dioxide	1050.0	15
sulphur oxides	3.0	3
nitrogen oxides	28.6	42
particulate matter	7.4	13
hydrocarbons	21.2	40
carbon monoxide	106.2	60

source (23)

major contributor to greenhouse gases; it generates about 18 per cent of all the carbon dioxide released from fossil fuels (8, 16, 17, 18), or about 15 per cent of total global anthropogenic carbon dioxide emissions. The chlorofluorocarbons contained in vehicle air-conditioning systems and upholstery are now being phased out (18).

Of all present day sources of noise, that from transport—above all from road vehicles—is the most diffused. In many countries it is the source that creates the greatest problems. Everywhere it is growing in intensity, spreading to areas until now unaffected, reaching ever further into the night hours and creating as much concern as any other type of pollution.

Recent data show that about 16 per cent of the population in OECD countries—approximately 110 million people—are exposed to road traffic noise in excess of 65 dBA, the level above which noise causes disturbance and harm (19, 20). For aircraft noise, about 0.5 per cent of the population in Europe and Japan are exposed to noise levels above 65 dBA, whereas the proportion of the population affected in the United States is 2 per cent. In many countries the percentage of population living in grey areas, that is those exposed to noise levels between 55 and 65 dBA, is increasing and therefore noise has become a more significant problem than it was a decade ago. The problem is becoming particularly severe in many urban centres in developing countries. Noise is a major problem in Bangkok, Cairo, Manila and many other cities.

Impact of transport on water

Oil pollution of inland waterways and of the marine environment results from normal discharges of barges and ships, and from accidental releases (Chapter 4, 9). In several developing countries, used motor oils are dumped on land or in

Noise from transport is growing in intensity, spreading to areas until now unaffected, reaching ever further into the night hours and creating as much concern as any other type of pollution.

surface water bodies, affecting the environment in a number of ways—for example, by polluting groundwater resources or damaging aquatic life. In some places used motor oils are discharged into sewers (especially at petrol stations and garages). This can create problems at sewage treatment plants by destroying or reducing the efficiency of micro-organisms that digest organic matter. Leakage from underground storage tanks, especially at petrol stations, has caused pollution of groundwater, for example, in the United States (21).

Responses

Over the past two decades, significant progress has been made in increasing the energy efficiency of new cars. New passenger cars in the United States are now almost twice as efficient as those of the early 1970s. On average, fuel consumption decreased from 16.6 litres/100 km in 1973 to 8.3 litres/100 km in 1987 (Figure 14.2). In the OECD countries, fuel consumption per car has decreased by an average of 25 per cent since 1970 (2, 7). This higher efficiency has been achieved mainly by weight reduction of cars, through substitution of steel by aluminium, plastic and ceramics in car manufacture, and by improvements in engines and transmission.

Progress has also been made in controlling vehicle emissions, especially in developed countries. Among the air pollutants, lead has been most successfully fought by phasing out leaded petrol. Between 1976 and 1987, lead in vehicle emissions dropped by 87 per cent in the United States (8). Similar results have been achieved in other OECD countries. However, in most developing countries leaded petrol is still the main fuel used.

Abatement of other vehicle emissions has been less successful and has been complicated by the fact that sometimes emissions of one pollutant can be lowered only by allowing levels of other pollutants to

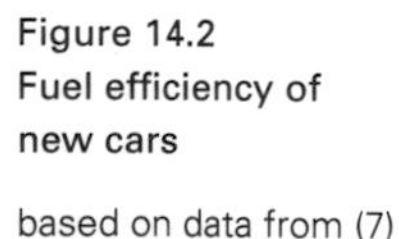
Figure 14.2
Fuel efficiency of new cars

based on data from (7)

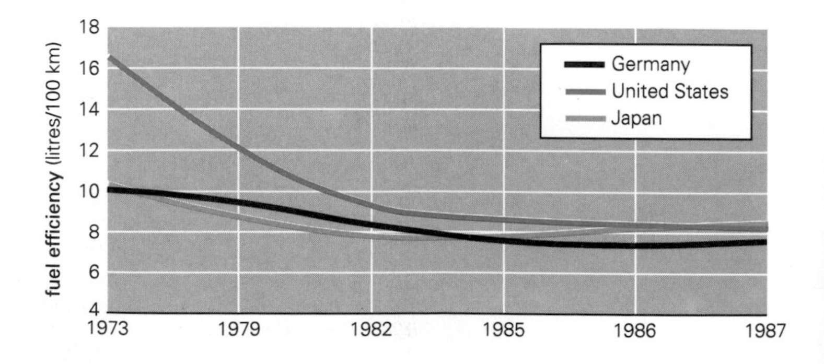

rise. For example, 'lean-run' engines (with an air to fuel ratio of 20 to 1 or more instead of the conventional 15 to 1) allow more efficient fuel combustion and reduce the emission of nitrogen oxides and carbon monoxide, but tend to increase emissions of hydrocarbons. And while a catalytic converter reduces carbon monoxide emissions, it slightly increases carbon dioxide and sulphur oxides emissions (8). Although significant declines in carbon monoxide and hydrocarbon emissions have been achieved in Canada, Japan, the United States and several other OECD countries, vehicle emissions are on the rise because of the increasing number of vehicles (6). This is particularly true in developing countries, where control measures are rarely implemented because of technical and economic problems (such as old vehicles, especially public vehicles and buses, inadequate maintenance and repair, and traffic congestion).

Regulatory measures (such as control of emissions, noise control, safety improvements of road, sea and air transport, traffic improvements and reduction of noise around airports) introduced in the past two decades, especially in developed countries, have helped reduce the impact of the transport sector on the environment. The fact that public passenger transport systems are more energy efficient and less polluting on a passenger-km basis (Figure 14.3) has been brought into focus. It has been demonstrated that the switch to buses and fixed rail transport systems for intracity travel in some countries has led to marked energy savings and a reduction in pollution (22). Reduction of highway speed limits has also led to an increase in fuel efficiency, increased vehicle tyre life, and has reduced road accidents in several countries.

Figure 14.3
Energy consumption and emissions of different city modes of transport

based on data from (24)

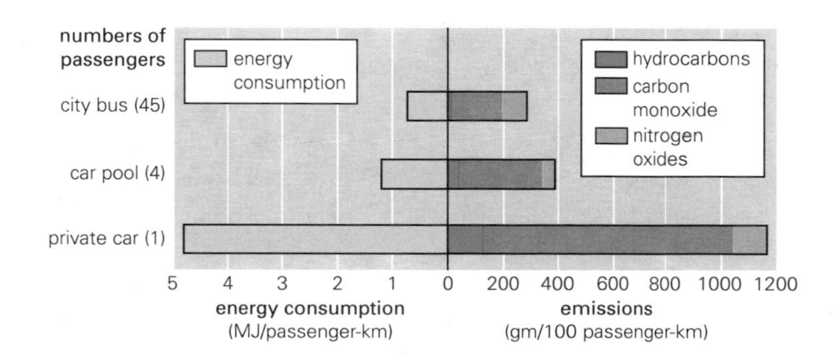

Chapter 15

Tourism

Tourism is big business. It has become a major worldwide industry and is expected to show continued strong growth. In the past two decades international tourist arrivals grew nearly threefold, and international tourist receipts rose from approximately US$22 billion in 1970 to about US$300 billion in 1990 (Figure 15.1). If domestic tourism and travel were included, these figures would be much higher. According to a study carried out for American Express Travel (1), travel and tourism accounted for sales of about US$1916 billion in 1987, making it the largest source of employment in the world. Over the past two decades, the bulk of tourism went to Western Europe, North America, Eastern Europe and the Soviet Union, and East Asia and the Pacific (Figure 15.2). The Mediterranean region accounted for an average of 36 per cent of international tourism (2).

Tourism expenditures as a contribution to gross domestic product (GDP) varies widely from one country to another depending on the size of the economy and level of expenditure. Many Caribbean states have tourism shares of GDP of 15 to 30 per cent. Although international tourism is often thought of as an easy means of contributing to the economic growth of developing countries, studies over the past two decades have demonstrated that the cost of the supplies and infrastructure needed for tourism have been very high in terms of foreign exchange; global balance sheets often show that many years must elapse before the receipt of the first real foreign exchange earnings from tourism-related activities. In fact, the balance of foreign exchange accruing to developing countries is relatively small (3).

In many developing countries, a significant part of the foreign exchange pays for the cost of importing goods and services used by tourists, some of the costs of capital investment in tourist amenities,

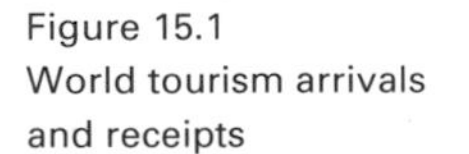
Figure 15.1
World tourism arrivals and receipts

based on data from various WTO reports

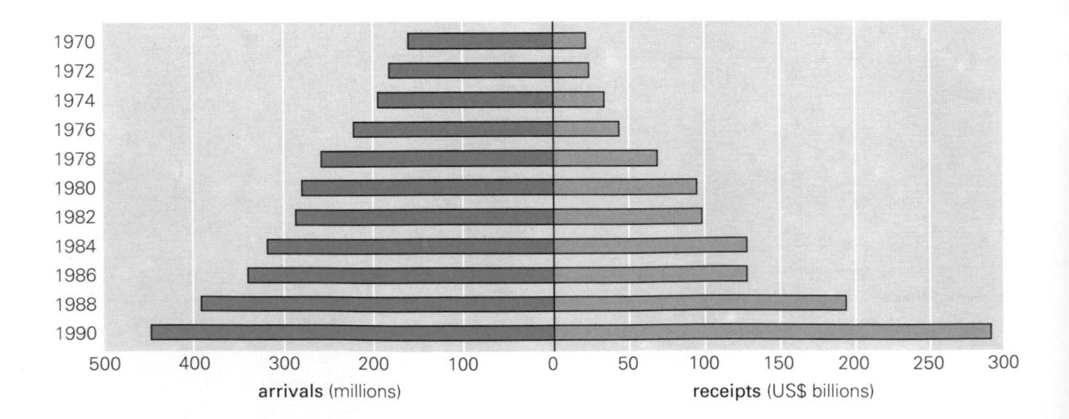

such as hotels and vehicles, payments to foreign travel agents, royalties, and promotion and publicity expenditure abroad. International tourist receipts are not, therefore, an indicator of the real income from tourism. The net income will vary from one country to another depending on the sums spent for tourist services and investment in the sector. It is now becoming increasingly clear that it is not tourism that leads to development, but a country's general development that makes tourism profitable.

Impact of tourism on the environment

Like other sectors of development, tourism can have both positive and negative impacts on the human environment. Tourism has positively benefited the environment by stimulating measures to protect physical features of the environment, historic sites and monuments, and wildlife. Recreation and tourism are normally the primary objectives of establishing and developing national parks and many other types of protected areas. These natural areas are becoming major attractions, and constitute the basis for what is now known as 'nature tourism' or 'ecotourism'.

Two main types of ecotourism exist: marine-based, and big-game/safari tourism. Case studies (at such sites as the Khao Yai National Park, Thailand, the Virgin Islands National Park, Virgin Islands, the Kangaroo Island, South Australia, and the wildlife parks in East Africa) have demonstrated that ecotourism yields direct financial benefits that outstrip the cost of maintenance and development of the parks. In addition, ecotourism stimulates employment and rural development in surrounding areas (4). The public in such areas is

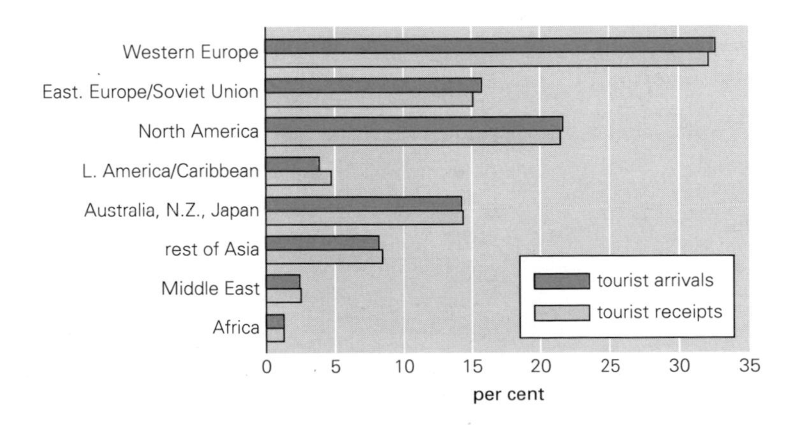

Figure 15.2
Tourist arrivals and receipts by region (1989)

based on data from (12)

becoming aware that environmental protection increases their economic gains, by increasing the number of visitors. In Tanzania and Rwanda, surveys have indicated that national parks should be protected and developed as long as they attract more tourists (5).

The historic and cultural heritage that determines the attractiveness of a country to tourists encourages the authorities to protect it. There are many examples of cultural salvage operations stimulated by tourism, and many efforts have been made to provide for the systematic protection of old towns, villages and groups of buildings of historic and artistic interest. UNESCO has supported many of these activities.

Tourism has been a driving force in establishing and improving summer and winter tourist settlements and health resorts. At Ixtapa, on the Pacific Coast of Mexico, a new tourist resort has benefited the neighbouring environment by providing such facilities as water supply, sewage systems, roads, electricity and tele- communications. Projects have been undertaken in many developing countries that have contributed both environmentally and economically to the improvement of the quality of life of local populations. 'Farm' or 'rural' tourism has grown in some countries—for example, the United Kingdom and France—contributing to the enhancement of farming and development in the countryside, and thereby discouraging excessive rural-urban migration (6, 7).

The environment, natural and anthropogenic, constitutes the basic asset of the tourist industry. If the carrying capacity of this asset is exceeded, it can deteriorate and may even be irreversibly damaged. Mass tourism (especially 'sand-and-sun tourism') has produced environmental damage of this kind in the Caribbean, Mediterranean and other seaside areas. In Barbados, the growing numbers of tourists have exerted increasing pressures on land use and infrastructure in the island. The increase of sewage discharges into the sea reduced the physical size of near-shore marine habitats. Water and electricity shortages have become common in Barbados, Grenada and Antigua because the carrying capacity of these services has been exceeded (8). In Tunisia the groundwater level in the Hammamet region has been lowered as a result of excessive withdrawal to meet the increasing needs of tourism (2). In Egypt, increased tourism has put excessive pressures on electricity consumption. One study (9) showed that one multinational tourist hotel in Cairo consumes sufficient electricity to meet the needs of 3600 middle-income households.

Tourism and recreation have affected coastal areas in a number of ways. The damage to coral reefs in Kenya, Tanzania, Madagascar,

Mauritius, Seychelles, Thailand , Malaysia and other countries has been well documented (10). The pressures of tourism on coastal areas is best illustrated by the situation in the Mediterranean region, which attracts about 36 per cent of international tourism (and much more of resident recreation). Pollution of coastal waters as a result of the increased discharge of sewage into the sea in high season has become a chronic phenomenon. Many countries (including Italy, France and Greece) had to close some beaches temporarily because the quality of their waters was not acceptable for bathing. A survey in the early 1980s of 1200 beaches in France showed that 30 per cent were not suitable for bathing (11). Similar numbers has been recorded in other countries.

Excessive tourism has created seasonal atmospheric pollution in some areas, including Spain, France and Italy. In Yugoslavia, where 86 per cent of international tourists arrive by road, seasonal atmospheric pollution due to tourism is the highest in the Mediterranean region (2). Syria, Turkey and Morocco are also increasingly affected by this seasonal increase in atmospheric pollution.

The increasing number of visitors to archaeological and historical sites is also a matter of concern: important sites can become trampled down, and in confined or underground areas artificial lighting, and even the visitors' breath, can have damaging effects. These problems are becoming acute in places such as Luxor in Egypt and Venice in Italy—as they are in many museums and art galleries (2, 8).

While tourism plays a major role in mountain area economies, ecosystem damage has in some instances reached a critical level, thus impairing the future of tourism itself. About 150 million visitor-nights are spent each year in the European Alps, and in high season local and tourist population density may reach a high of 1800 people/km^2, higher than that of many industrialized districts. Such excessive pressure affects the mountain ecosystem: the soil, vegetation, wildlife and water balance. The Mount Everest region in Nepal, once very isolated and rarely visited, has become a victim of success. Now there is a major trekking and climbing industry in the area. Major management problems include garbage and waste disposal, and excessive firewood collection (4).

The Mount Everest region in Nepal, once very isolated and rarely visited, has become a victim of success. Now there is a major trekking and climbing industry in the area.

Responses

The relationship between tourism and environment is one of a delicate balance between development and safeguarding the environment. The Manila Declaration (1980) emphasized that the needs of tourism must not be satisfied in a fashion prejudicial to the social and economic interests of the population in tourist areas, to the environment or, above all, to natural resources and historical and cultural sites, which are the fundamental attraction for tourism. It stressed that these resources are part of the heritage of mankind, and national communities and the international community must take the necessary steps to ensure their preservation. Long-term and environmentally sound planning is a pre-requisite for maintaining a balance between tourism and environment, and for ensuring that tourism is a sustainable development activity.

Burdened by foreign debt and desperate for hard currency, many developing countries have, however, shrugged off their worries that tourism could degrade the natural environment—the beautiful resource that makes them attractive. These short-sighted policies have, in fact, led to a marked degradation of the environment in some countries, which now keeps many tourists away. It would take years and massive financial resources to redress this degradation and re-accelerate tourism again. On the other hand, efforts have been made in many countries to establish and/or develop protected areas, and protect wildlife (Chapter 8). Efforts to improve the environment in coastal zones are also under way in several countries (Chapter 4).

Short-sighted policies have led to a marked degradation of the environment in some countries, which now keeps many tourists away. It would take years and massive financial resources to redress this degradation and re-accelerate tourism again.

Part III

Human Conditions and Well-being

Chapter 16

Population growth and human development

Between 1970 and 1990, the world population grew by 1.6 billion; of that growth, 90 per cent occurred in developing countries (Figure 16.1). In the next two decades, it is projected that the world's population will grow by 1.7 billion, reaching about 7 billion in the year 2010. Though population growth rates have been falling steadily since 1970, both in developed and developing regions (Figure 16.2), the net annual addition to the number of people in the world has been rising since the 1970s: the 1990s will witness the largest average annual increment to world population in history (Figure 16.3, on page 180). After this, population growth should slow down and by 2110 the world population may reach a stationary level of 10.5 billion (1, 2).

The average annual rate of population growth in developed countries decreased from 0.86 per cent a year in the period 1970–75 to 0.53 per cent a year in the period 1985–90. By contrast, the annual rate of population growth in developing countries decreased from 2.38 per cent a year during 1970–75 to 2.10 per cent a year during 1975–80, and has since remained constant. However, this overall trend was subject to regional differences. In East Asia, South-East Asia, Central America and the Caribbean, there were marked declines in population growth rates in the 1980s. But in Africa the growth rate actually increased over the past decade, and is now estimated at 3 per cent a year. In Asia, population growth rates differ significantly from one sub-region to another. China, with nearly a fifth of the world's population, has dramatically reduced its population growth rate in recent years—from 2.20 per cent a year in 1970–75 to 1.23 per cent per year in 1980–85; however, there was a slight increase to 1.39 per

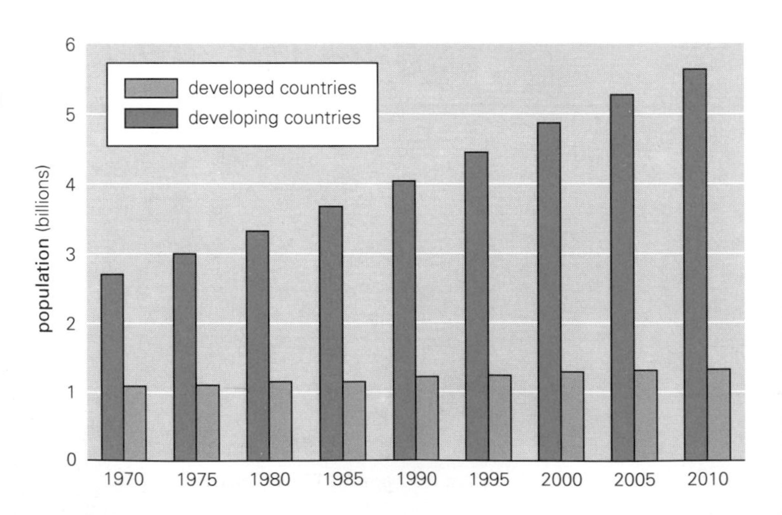

Figure 16.1
World population 1970–2010 in developed and developing countries

based on data from (2)

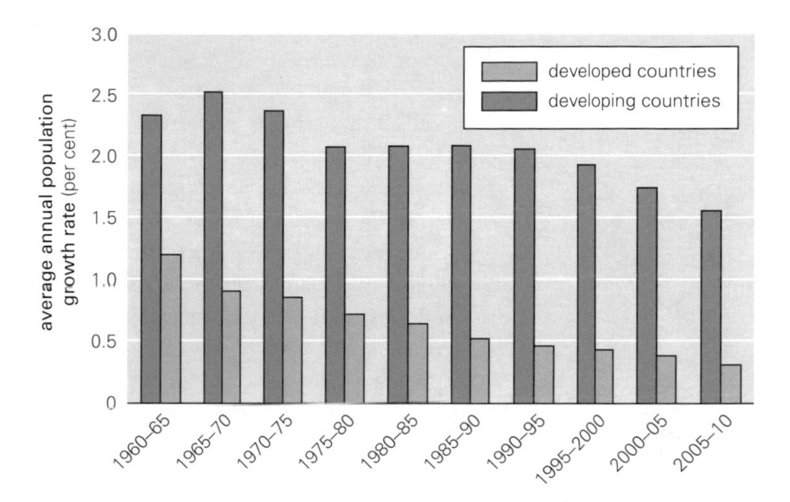

Figure 16.2 Average annual population growth rates in five-year periods from 1965–2010

based on data from (2)

cent a year in the period 1985–90 (2). The world's five most populous countries, China, India, the Soviet Union, the United States and Indonesia, accounted for 51 per cent of the world's population in 1990 and will account for half of the world's population in 2000. The same five countries will account for 42 per cent of the total world population growth between 1990 and 2000.

While birth and death rates have fallen worldwide (Figure 16.4), life expectancy at birth rose from an average of 56.7 years in 1970–75 to 61.5 years in 1985-90, and is projected to increase further in future (2). The infant mortality rate fell from 94 per 1000 births a year in 1970–75 to 71 per 1000 births per year in 1985–90.

Yet enormous gaps remain between the rich and poor in both developed and developing countries, and between the developed and developing regions. Life expectancy at birth now exceeds 73 years in developed countries, compared to 60 years in developing countries. Differences also exist between developing regions. In Africa, life expectancy at birth is only 52 years while in South Asia it is 57 years and in Latin America 66 years. Infant mortality rates have fallen in nearly 150 countries in the past decade, and industrial countries now have the lowest infant mortality rate (9 per 1000 live births). By contrast, the infant mortality rate is still more than 100 per 1000 live births in

In the next two decades, it is projected that the world's population will grow by 1.7 billion, reaching about 7 billion in the year 2010. Though population growth rates have been falling steadily since 1970 ... the 1990s will witness the largest average annual increment to world population in history.

34 developing countries—2 countries in Latin America and the Caribbean, 2 in the Middle East and North Africa, 23 in Africa south of the Sahara, and 7 in Asia (3).

Family planning is essential in population control. Though the elements of family planning are well understood, questions remain on how to promote and deliver it effectively. Early marriage and child-bearing in the developing countries are inextricably tied to the economic and social rewards that societies attach to children. The experience of the industrialized world has shown that development is the best means towards population control. Population growth, development and a productive environment form three points of an interlinked triangle: progress cannot be made in any one area unless progress is made in the other two. The World Fertility Survey has found that women would have an average of 1.41 fewer children if they were able to choose the size of their family. In as little as 35 years from now this could make a difference of approximately 1.3 billion people to the world's population (3). Therefore, without the provision of adequate women's education and without radical improvements in the status of women, family planning cannot fully succeed. And this cannot be achieved without development.

The most telling and tragic indicator of poverty is a high infant mortality rate. In poor societies, experience has repeatedly demonstrated that attempts to lower birth rates—and population growth—cannot be separated from efforts to keep children alive and healthy. The survival of children is one of the major motivating forces

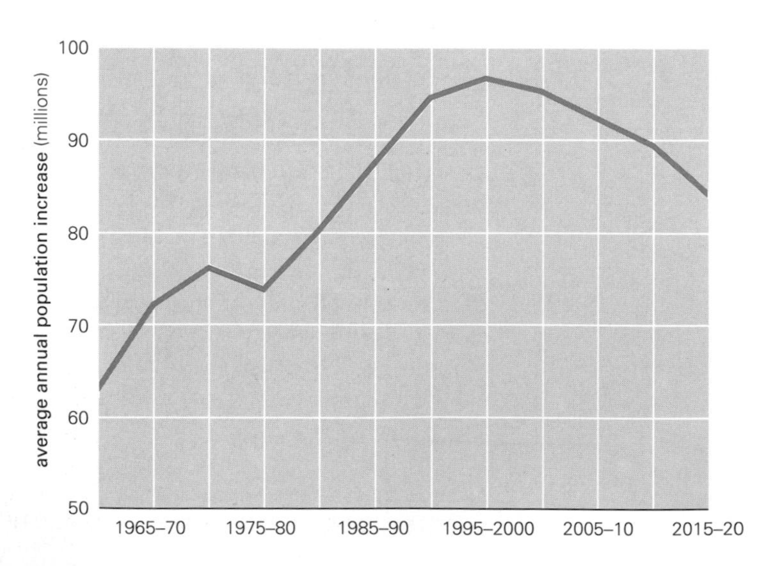

Figure 16.3
Average annual population increase worldwide 1965–2020

based on data from (2)

in the desire of parents for smaller families (4). Successful programmes to improve birth spacing revolve around a range of health and literacy initiatives. If these were implemented worldwide, global replacement level fertility rates (of slightly more than two children per couple) could be achieved by the year 2010; and the world population could stabilize at 7.7 billion by 2060. If, however, this replacement level fertility rate were not reached until 2065, the world population would reach 14.2 billion by 2100 (5). Such an explosion would obviously mean fewer resources per capita than are available today—a startling example of intergenerational irresponsibility and inequity.

People, resources, environment and development

An underlying theme of the past two decades has been the recognition that development is multidimensional, encompassing not only economic and social issues, but also those related to population, the use of natural resources and management of the environment (6). The growing focus on the interrelation between people, resources, environment and development has stemmed from three basic considerations. First, it has become increasingly evident that development at national and regional levels affects productive processes in a variety of ways, not all of them beneficial. Second, while such effects involve the vigorous interaction of economic, social, demographic and physical factors, it is difficult to establish their

Figure 16.4
Annual average birth and death rates per 1000 population

based on data from (2)

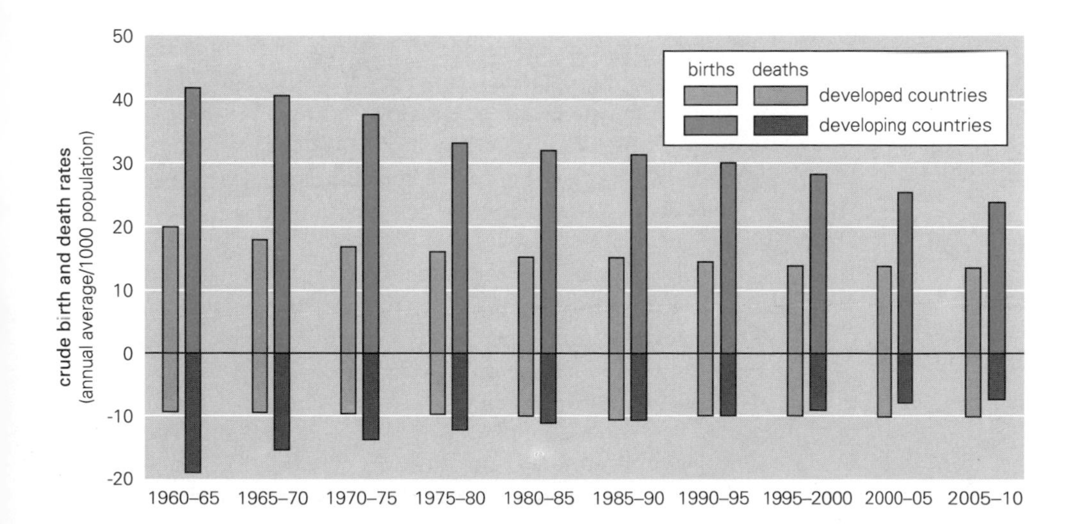

causal relationships. Third, there has therefore been great uncertainty about the likely long-term impact of development as it affects quality of life and environment; appraisals have tended to focus on risks of negative rather than positive outcomes.

The relationship between population, resources, environment and development is very complex and all these factors interact in different ways in different places. The pace of development, its content, location and the distribution of its benefits, largely determine the state of the environment. Such factors also influence population growth and distribution. Environmental resources provide the basis for development, and environmental benefits brought about by development can improve the quality of life. Similarly, the size of population, its growth rate and its distribution pattern can influence the state of the environment, just as they condition the pace and composition of development.

Many global and regional models have been used in the past two decades in attempts to determine future consumption of resources and their availability. These include the Global 2000 Model, the IIASA Energy Models, the Latin American World Model, the Mesarovic-Pestel World Model, the MOIRA Model of International Relations in Agriculture, the United Nations World Model, the Club of Rome's Worlds 2 and 3, and several others. Inherent uncertainties in the assumptions made in these models, and the limited number of factors that can be incorporated in them, mean that they can provide only indicative results. The models have, however, been useful in identifying gaps in knowledge. The construction of a single aggregate global or regional model that incorporates all the variables in the equation of population-resources-environment-development remains a challenge to the scientific community. The question posed at the time of the Stockholm Conference in 1972 is still as valid—and without answer—today as it was then: 'Is there any way to meet the needs and aspirations of the 5 billion people now living on the earth without compromising the ability of tomorrow's 8–10 billion to meet theirs?'

Population growth need not necessarily lower standards of living, impair the quality of life or cause environmental degradation. Global and historical assessments of the earth's capacity and human ingenuity to produce goods and services have prompted some optimistic projections (7, 8). World population growth has, in the past, been accompanied by a steady increase in the world's capacity to provide for the necessities and amenities of human life. The problem, therefore, is not simply one of numbers: it is also the widening disparity in consumption patterns and lifestyles between the rich and the poor. A child born in a rich, industrialized country, or in a rich family in a developing country, where per capita consumption of energy and materials is high, places a much greater burden on the planet than a child born in a poor country (9). Two groups in particular are responsible for a disproportionate share in the consumption of resources and in environmental degradation: the world's billion richest and billion poorest people (10). Those at the top consume the largest slice of the earth's resources, and generate enormous quantities of waste. Those at the bottom have the highest fertility rates and in their fight for survival are responsible for a disproportionate amount of environmental destruction.

Human development

In the past two decades, several indices have been proposed for measuring the quality of life—these include the physical quality of life index (11), the human suffering index (12) and, more recently, the human development index introduced by UNDP (13, 14). These indices have brought into focus the widening gap between the North and South. The developing countries, with 77 per cent of the world's population, earn only 15 per cent of world's income. The average GNP per capita in the North (US$12 510) is now 18 times the average in the South (US$710). According to the human development index, about 2 billion people are at the lowest level of development (14), and most of these are the poorest in the world.

Poverty has been defined in a number of ways, but perhaps the most eloquent definition is by Robert McNamara, the former president of the World Bank. He described absolute poverty as 'a condition of life so limited by malnutrition, illiteracy, disease, squalid surroundings, high infant mortality, and low life expectancy as to be beneath any reasonable definition of human decency' (15). This shows that poverty is far more than just an economic condition. The World Bank (16) has recently used two poverty lines to estimate the number of poor people in developing countries. Those whose annual consumption is less than US$370 per person a year are considered poor, and those whose annual consumption is less than US$275 per person a

A child born in a rich, industrialized country, or in a rich family in a developing country, where per capita consumption of energy and materials is high, places a much greater burden on the planet than a child born in a poor country.

year are considered extremely poor. According to these definitions, there are an estimated 1116 million poor people in developing countries, of whom 630 million are extremely poor. The UNDP (14) estimates that the number of poor people in the developing countries will rise to 1.3 billion by 2000 and probably 1.5 billion by 2025.

These figures conceal considerable variations within and among countries. The burden of poverty is, in fact, spread unevenly between regions in the developing world, among countries within those regions, and between localities within those countries. Nearly half of the world's poor live in South Asia, but there is also a heavy concentration of poverty in Africa. It is estimated that Africa's share of the world's poor will rise from 30 per cent today to 40 per cent by 2000, overtaking Asia (14).

Global economic performance over the past two decades has been erratic and has varied widely among countries and continents. In general, there has been an economic deterioration in much of the developing world. These two decades have seen escalating external debts, falling prices for raw commodities, and adjustment policies that have weighed heavily on the poor. Living standards for millions in Latin America are now lower than in the early 1970s; in most of sub-Saharan Africa, living standards have fallen to levels last seen in the 1960s (13).

As individual economies have foundered in the inhospitable world economy of the past two decades, many developing countries have pursued structural adjustment policies. These have usually taken the form of dampening demand, devaluing currency, withdrawing subsidies on fuel and staple foods, and severely cutting government spending. But in spite of these measures, economic recovery and structural change have been slow in coming. Furthermore, the impact of income declines and social services cutbacks became manifest. Studies by UNICEF (3) showed that, in 37 poor nations, spending per capita on schools fell by about 25 per cent in the 1980s. Health spending per capita has declined in more than three-quarters of African and Latin American nations. In several countries in Latin America and sub-Saharan Africa, the historical decline in infant mortality has been reversed and the incidence of malnutrition has increased. The basic problem in applying structural adjustment has been that insufficient attention has been paid to its effects on the poor; accumulating evidence suggests that many structural adjustment measures have damaged the poor disproportionately. By the end of the 1980s, the issue of adjustment had come under scrutiny by many agencies (3).

The problems of developing countries have been compounded by a dramatic increase in their foreign debt, which now totals more than US$1.3 trillion—requiring nearly US$200 billion a year in debt servicing alone (13,14). There have also been important changes in the way resources are now moving across the North-South boundary. Before 1984, the net flow was progressive: industrial countries gave more to developing countries in loans and grants each year than they took back in interest and principal payments. But by 1990, the South was transferring at least US$20 billion a year back to the North. Reduced prices paid by industrialized nations for the developing world's raw materials have resulted in losses to developing countries equalling a further US$40 billion each year (17).

The dismal economic predicament of developing countries both causes and aggravates environmental despoliation. This in turn makes economic and structural reform difficult to achieve. It has long been recognized—particularly since the Stockholm Conference—that poverty is one of the greatest threats to the environment. The developing countries make many choices that lead to environmental degradation because of the imperative of immediate survival, not because of a lack of concern for the future. Economic deprivation and environmental degradation have thus come to reinforce one another and to perpetuate destitution in many developing countries. A high priority for the world community must be to agree on ways and means—many of which are already known—to stop this vicious cycle.

The dismal economic predicament of developing countries both causes and aggravates environmental despoliation. This in turn makes economic and structural reform difficult to achieve.

Chapter 17

Human settlements

A human settlement is a community or group of people living in one place (1). The development of such a community for productive purposes involves a transformation of the natural environment into a man-made environment that includes a variety of structures and institutions designed to meet the community's needs in work, recreation and other aspects of human life. It thus has a natural setting, a physical infrastructure of energy sources, housing, transport, water and waste disposal services; and a social infrastructure of political, educational and cultural services.

Throughout the world, the single most frequent form of human settlement is still the village. There are far fewer cities and towns than villages, isolated farmsteads and herding camps. In 1970, 62.9 per cent of the world's population lived in rural areas; in 1990 this proportion declined to 57.4 per cent and is expected to decline further to about 40 per cent by the year 2025 (2), mainly as a result of rural-urban migration.

Urbanization has been growing much faster in developing countries than in developed countries (Figure 17.1). In developing countries, the level of urbanization increased from 25.4 per cent in 1970 to 33.6 per cent in 1990; this level is expected to reach 39.3 per cent by the end of the century, and 57 per cent by the year 2025 (2). By the year 2000, 77 per cent of Latin America's population, 41 per cent of Africa's and 35 per cent of Asia's population will be urbanized. The urban population in the developing countries is growing by 3.6 per cent a year, whereas in industrialized regions the urban population is growing by only 0.8 per cent a year (3). The uncontrolled expansion of towns and cities in the developing countries has overwhelmed transport, communication, water supply, sanitation and energy systems. It has also created a vast array of environmental, social and economic problems.

Urbanization and environment

Despite technological achievements that enable people in developed and in some developing countries to live and work in high-rise buildings, the most common pattern of urban growth is still urban sprawl. This urban sprawl chews up land—in some countries valuable agricultural land. Between 1980 and the end of the century, urban areas in developing countries will more than double in size, from about 8 million hectares to more than 17 million hectares (3). In some countries this will mean a loss of land additional to that lost by over-use and mismanagement (see Chapter 6). On the one hand this could

Throughout the world, the single most frequent form of human settlement is still the village. There are far fewer cities and towns than villages, isolated farmsteads and herding camps. In 1970, 62.9 per cent of the world's population lived in rural areas.

lead to a further deterioration of rural areas. On the other, it could create more food supply problems for the urban areas which normally depend on nearby farms for supplies of agricultural products. Even in countries such as the United States and Australia, where food and raw material supplies for cities are transported over long distances, there are examples of urban expansion having harmful effects on specialized local rural production—such as wine growing around Adelaide in Australia, and fruit and vegetable production in California (1).

Urban growth can also have important impacts on neighbouring areas through the increased demand for natural resources, and the increase in the inflow and outflow of materials, products, energy, water, people and wastes. A study of fuelwood use in Kenya has shown that one of the major contributors to deforestation was the wholesale conversion of wood to charcoal for sale to people living in towns and cities (3). In several countries, for example in Egypt, there has been a growing tendency among farmers living near cities to grow the more profitable agricultural products—such as vegetables and fruits—needed by urban consumers at the expense of staple crops (4).

The increased movement of goods and people to and from urban areas has created a need for better transport which, with its associated infrastructures, entails the use of even greater land areas. Chronic traffic congestion in and around cities is a by-product of urban growth. In Bangkok, traffic jams are now so severe that the

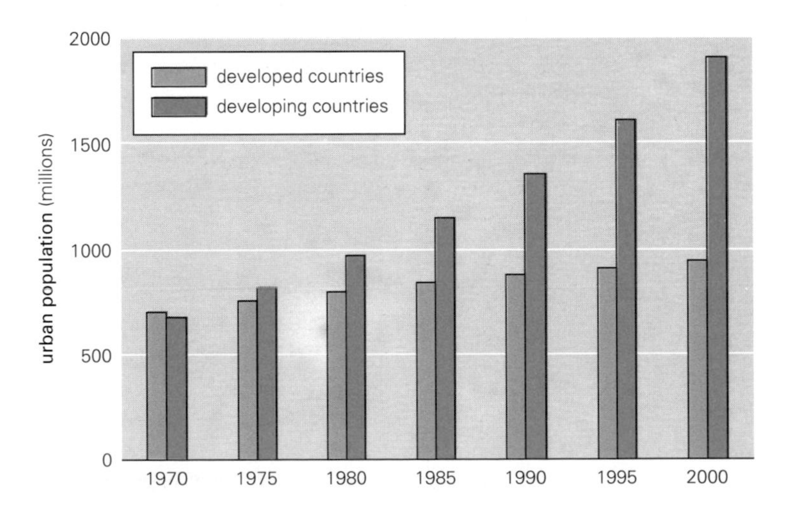

Figure 17.1
Number of urban dwellers in developed and developing countries 1970–2000

based on data from (2)

time lost by passengers on city streets plus the amount of extra petrol consumed are reckoned to cost at least US$1 billion a year. A further US$1 billion is lost in medical bills and through worker absenteeism due to ailments related to air pollution (3). Similar chronic congestion exists in most cities in developing countries.

As cities increase in size, slums and squatter settlements proliferate. It has been estimated that about one-third of the urban population in developing countries—about 200 million in 1970 and 450 million in 1990—live in urban slums and shanty towns. The percentage of people living in such areas varies markedly from one city to another, and from country to country (Figure 17.2). But most of them share the same precarious, dismal environment: over-crowded sub-standard shelters, insufficient and inadequate clean water supplies, lack of sanitation and refuse collection services, and lack of proper roads. Many of these shanty dwellers—sometimes referred to as the urban poor or those living on the margins (5)—are unemployed, uneducated, undernourished and chronically sick. Socio-economic and environmental conditions in the slums are best illustrated by intra-urban differentials in health. In Manila, the infant mortality rate for the whole city was 76 per 1000, compared with 210 per 1000 in Tondo, a squatter area. Neonatal mortality in Manila was 40 per 1000, but it was 105 per 1000 in Tondo (5). In Buenos Aires mortality due to tuberculosis was three times higher in peripheral areas than the average for the city as a whole. Similar studies exist which point to intra-urban differentials in morbidity. A greater prevalence of

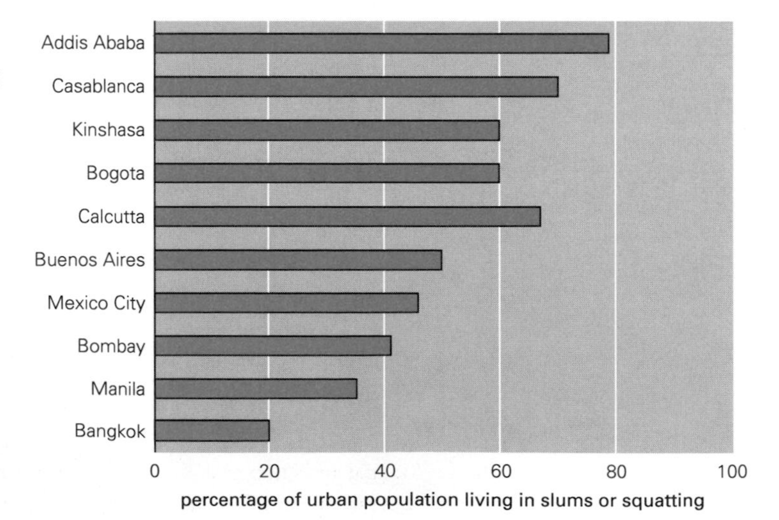

Figure 17.2
Percentage of people living in slum areas and as squatters in selected urban centres

based on data from (6)

diarrhoeal diseases and various infections is associated with poor housing, water and sanitation facilities. In addition, those living in slums and squatter settlements are more prone to natural hazards and the impacts of industrial accidents (see Chapter 9).

Unplanned urban growth has caused an acute housing shortage in many countries. In developing countries, the percentage of households unable to afford dwellings of normal standard in selected cities—such as Cairo, Manila and Bangkok—has increased over the past two decades from 35 to 75 per cent. The result has been increased overcrowding, a proliferation of substandard housing and squatter settlements. The average housing occupancy rate in developing countries is now about 2.4 persons per habitable room, compared to 0.8 in developed countries (7).

Although water supplies to urban areas in developing countries have improved over the past two decades, in 1990 there were still about 244 million people—18 per cent of the urban population—without clean water supplies (see Chapter 5), compared to 33 per cent in 1970. The provision of sanitary facilities has not improved over the past two decades. In 1990, 377 million people in urban areas, or 28 per cent of the population, did not have adequate sanitary facilities; in 1970 that figure was 29 per cent. In fact, most urban centres in Africa and Asia have no sewage system at all—including many cities with a million or more inhabitants (8). Most human and household wastes end up untreated in rivers, streams, canals, gullies and ditches. In those cities that do have sewerage, rarely are more than a small proportion of the population served—typically those living in richer, residential, government and commercial areas.

Daily per capita domestic refuse generation in cities in developed countries is estimated to vary between 0.7–1.8 kg, whereas the figure is somewhere between 0.4–0.9 kg in developing countries (9). On average, the amount of municipal solid wastes generated in the developed countries increased from 318 million tonnes in 1970 to 400 million tonnes in 1990, an increase of about 25 per cent. In developing countries, there were about 160 million tonnes of refuse in 1970; this doubled to 322 million tonnes in 1990.

Refuse collection services are inadequate or non-existent in most residential areas in Third World cities: an estimated 30–50

About one-third of the urban population in developing countries live in urban slums and shanty towns. Most share the same precarious, dismal environment: over-crowded sub-standard shelters, inadequate clean water supplies, lack of sanitation and refuse collection services, and lack of proper roads.

per cent of solid wastes generated within urban centres is left uncollected. It accumulates on streets, on open spaces between houses, and on wasteland. Such untreated refuse, particularly in hot climates, creates a breeding ground for disease vectors and pathogens. Where municipal solid wastes are managed, hand picking of refuse is the most viable economic option. Crude dumping is almost universal in developing countries and the practice often supports a large population of scavengers who earn their daily bread by extracting materials from the waste and selling them. Paradoxically, the poorest countries are thus achieving a high level of recycling despite the small proportion of saleable matter in the waste. As whole families are usually employed in scavenging, the grim realities of child labour and associated public health hazards should not be overlooked. The young and the elderly are exposed to a wide variety of pollution effects, obnoxious odours and, especially, diseases that may seriously endanger the health of all people who come in direct contact with the waste (10).

Though several technologies for municipal solid waste management are available, the problem of adequate waste disposal involves more than technology. Today, social and political considerations—such as public participation, particularly in the process of decision making, and recognition of the important economic role of scavengers—are influencing solid waste management and may play a greater role than technical innovation in bringing about future beneficial changes (11).

Rural settlements

Conditions for people living in rural areas are in general no better today than they were in 1970. Many houses are still well below standard, made of mud bricks, bamboo, wood and other locally-available material. Though the percentage of the rural population served with clean drinking water increased from 14 per cent in 1970 to 63 per cent in 1990, there were still 988 million people without access to clean water supplies (see Chapter 5). The proportion of the rural population served with some sort of sanitary facilities increased from 11 per cent in 1970 to 49 per cent in 1990, but there were still about 1364 million people without any such facilities. In rural areas, obtaining water and making it more readily available for domestic use has traditionally been women's work. In many developing countries, women (and children) still have to walk long distances to bring home fresh water. Such water is mainly used for cooking and drinking; canal or pond water is still generally used for washing and bathing, especially of children, in many rural areas. Wood, agricultural residues and cow dung are still the main sources of fuel in rural areas. Again, women and children are responsible for collecting branches, bushes, crop residues and cow dung. Electricity is still a rare commodity in most rural homes, in spite of efforts to increase rural electrification.

Human settlements and health

The environment in and around human dwellings offers an important habitat for a wide range of insects and rodents. Overcrowded, substandard housing—with inadequate water supplies and sanitation facilities—offers fertile areas in which fleas, cockroaches, bugs, mosquitoes, flies, other insects, rats and rodents can flourish. These insects transmit a variety of diseases (12), the best known of which is Chagas disease, transmitted by bugs that live in cracks and crevices of poor-quality houses in Latin America. According to WHO estimates, about 500 000 people become infected every year, 300 000 of them children. Between 10 and 15 per cent of infected people die during the fever typical of the acute phase of the disease (13). The rest become chronically infected, and ultimately suffer heart and other chronic disorders. The WHO estimates that 16–18 million people in South America are infected and another 90 million are at risk. A programme has been recently launched to eradicate Chagas disease from parts of Argentina, Brazil, Bolivia, Chile, Paraguay, southern Peru and Uruguay (13). Other diseases characteristic of poor human settlement conditions include filariasis, malaria, typhoid, dengue and yellow fever (see Chapter 18).

The use of wood, agricultural residues, coal and dung for domestic fires and other uses in rural areas creates severe indoor air pollution, to which women and children are particularly exposed. Studies carried out in the 1980s (14, 15, 16) provided evidence of the increase of incidence of respiratory diseases and naso-pharyngeal cancer among persons exposed to emissions of such fuels in rural homes.

Because many coastal cities discharge their sewage into the sea without treatment, coastal bathing water can be hazardous to health. Studies carried out in Canada, Egypt, France, Hong Kong, Israel, Spain and the United States showed a high incidence of eye

Human dwellings offer an important habitat for a wide range of insects and rodents. Overcrowded, substandard housing—with inadequate water supplies and sanitation facilities—offers fertile areas in which fleas, cockroaches, bugs, mosquitoes, flies, other insects, rats and rodents can flourish.

infections, skin complaints, gastrointestinal symptoms, and ear, nose and throat infections due to exposure to polluted bathing water (17). It has been estimated that 40 per cent of tourists on vacation at Mediterranean coastal resorts become ill during or immediately after their visit (see Chapter 15). The discharge of industrial waste into the sea creates additional hazards that impairs not only health but also the environment around coastal cities (see Chapters 4 and 10).

Responses

The problems and opportunities provided by human settlements differ in size and kind between developed and developing countries. How far advanced developed and developing countries are in human settlements policies is difficult to judge. But the conditions outlined here, including widening intra-urban and rural-urban differences, indicate inadequate responses in most countries to existing and emerging problems. In developing countries, in particular, governments have not been able to cope with the increased demands for infrastructures and services that have accompanied massive growth in urbanization. Economic burdens carried by developing countries have hampered investment in proper rural and urban development.

During the past two decades it has become evident that conventional mechanisms for financing the housing of low-income families could not resolve their difficulties; and that rent control was a weak tool. It has also become evident that clearing squatter areas will not solve the problems of poor housing (18). Grassroots initiatives have too often been thwarted by institutional arrangements and government policies. Innovative efforts to improve the situation have been made in countries such as Chile, the Dominican Republic, El Salvador and the Philippines. The need for much broader assistance targeted to reach lower-income groups has been generally recognized.

How far advanced developed and developing countries are in human settlements policies is difficult to judge. But the conditions outlined here ... indicate inadequate responses in most countries to existing and emerging problems.

Chapter 18

Human health

All constituents of the environment of our planet ultimately exert an influence on human health and well-being. However, the greatest and most direct influence is the immediate environment of home, work place and neighbourhood. Both environmental and genetic factors are involved in the incidence of disease. While genetic factors usually give rise to congenital diseases and environmental factors to acquired ones, there is often an interplay between the two.

Freedom from organic disease is usually considered synonymous with a reasonable state of health, but freedom from non-organic disease is also important. Good health demands a sound mind in a sound body. The socio-economic implications of impaired mental health in any population group cannot be ignored. Impaired mental health, like its organic counterpart, can be caused by genetic or environmental factors, or an interplay between both. Over the past two decades, evidence of the role of biochemical change in causing mental ill health has increased. Some of these biochemical abnormalities could be inherited and some environmentally induced. Certain organic causes of mental ill health are found in environmental factors—as with the group of psychoses resulting from trypanosomiasis and other infectious agents. Exposure to heavy metals such as mercury or lead and to certain synthetic compounds may create a predisposition to brain tumours or abnormal behaviour. A study on the long-term effects of exposure to low doses of lead in childhood indicated that such exposure is associated with deficiencies in central nervous system functioning that persist into young adulthood (1).

Figure 18.1
Prevalence of anaemia worldwide in the early 1980s

based on data from (2)

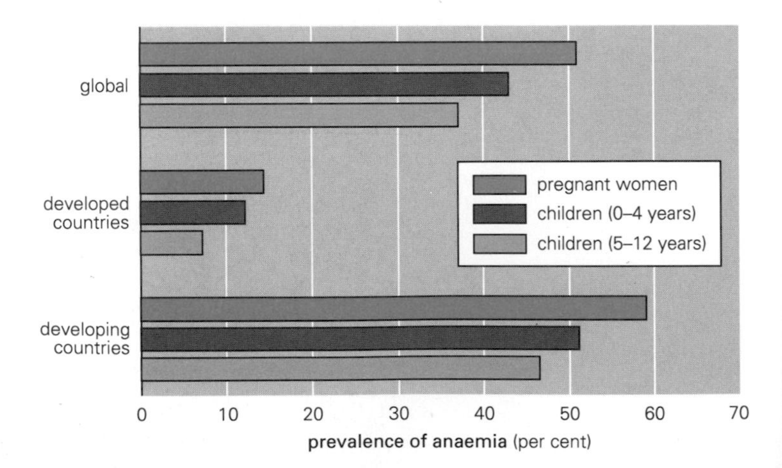

Malnutrition is the most pervasive cause of ill health ... and a major contributor to the high death rate among infants and young children in developing countries.

Malnutrition

Malnutrition is the most pervasive cause of ill health (see Chapter 11), and a major contributor to the high death rate among infants and young children in developing countries. An infant's birth weight is the single most important determinant of its early chances of survival, healthy growth and development. Because birth weight is conditioned by the health and nutritional status of the mother, the proportion of infants with a low birth weight (less than 2500 g) accurately reflects the health and social status of women and of the communities into which children are born. In communities where malnutrition is a chronic problem, or during periods of food shortages or of physical stress such as recurrent droughts, pregnant women rarely get enough to eat and foetal growth suffers. Approximately 51 per cent of pregnant women in the world suffer from nutritional anaemia (low haemoglobin levels due to poor diet); the percentage in developing countries is 59, much higher than the 14 per cent encountered in industrialized countries (Figure 18.1). Some 22 million (about 16 per cent) of the 140 million infants born each year in the world have a low birth weight. At least 20 million of these infants are born in developing countries, the majority (more than 13 million) in South Asia and the rest in Africa, Latin America and East Asia (3, 4). The Global Strategy of Health for All, launched by the World Health Assembly, aims at achieving a target birth weight of at least 2500 g for 90 per cent of newborn infants, and adequate growth of children, measured by weight-for-age goals, by the year 2000 (4).

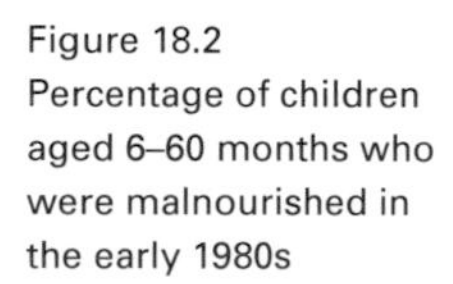

Figure 18.2
Percentage of children aged 6–60 months who were malnourished in the early 1980s

based on data from (11)

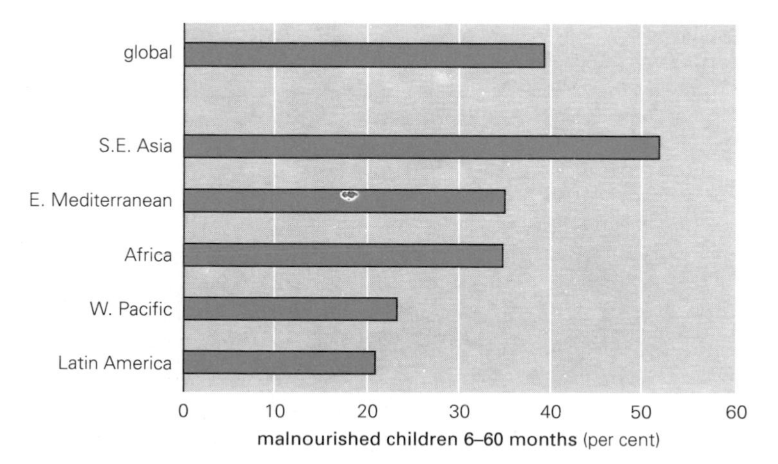

In terms of numbers affected, malnutrition is the most serious condition affecting the health of children, particularly in developing countries. Surveys in different regions of the world indicate that at any moment an estimated 10 million children are suffering severe malnutrition, and a further 200 million are inadequately nourished (Figure 18.2). Malnutrition makes a child (or an adult) more prone to infection, and infection may exacerbate the effects of malnutrition.

The best protection for infants against both malnutrition and infection is breast feeding and the past two decades saw a heightened awareness of its importance. However, nearly all chemical compounds ingested by the mother will be found in her milk in one form or another. Cadmium, lead, mercury, DDT, its derivatives and other pesticides have all been found in human milk in several countries. Studies (5, 6, 7, 8) have revealed that in some countries, concentrations of DDT and DDE in human milk are higher than acceptable daily intake criteria and maximum residue limits established by WHO and FAO. However, no evidence has been found to suggest that the levels of DDT and DDE generally found in human milk have harmed infants. In fact, breast milk tends to be much less contaminated than its substitutes. The prevalent high death and disease rates among artificially-fed infants in many developing countries can be attributed to improper preparation, as well as contamination, of formula milk and other infant foods. The acceptance of WHO's international code for marketing of breast milk substitutes has stimulated creation of energetic government-designed programmes to promote breast-feeding. Despite the increasing popularity of breast-

Figure 18.3
Main causes of death in developed and developing countries in the mid-1980s

based on data from (10)

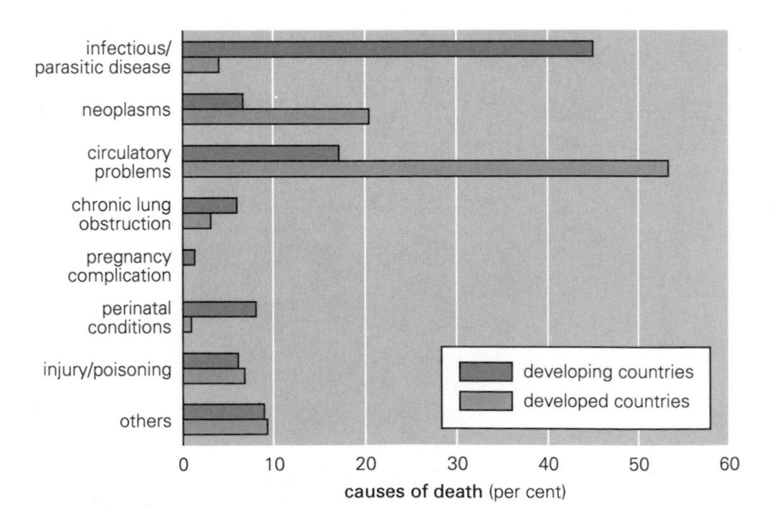

feeding in the industrialized countries, no similar increase has occurred in developing countries (9).

Communicable diseases

Different environmental conditions determine seasonal as well as regional differences in the incidence of disease. Figure 18.3 illustrates the main causes of death in the mid-1980s (10). Infectious, parasitic, perinatal and pregnancy complications predominate in developing countries. Some communicable diseases are transmitted much more easily during the rainy season. Temperature, humidity, soil, rainfall and atmospheric conditions are all important factors in the ecology of certain infective and infectious diseases, especially because they control the distribution and abundance of their vectors.

Communicable diseases account for a large proportion of illness and death in developing countries, where billions of people still lack such basic necessities as adequate shelter, access to safe water supplies and sanitation, and refuse disposal facilities. The deteriorating environmental conditions in which people live propagate the spread of infective agents and the breeding of their vectors. Overcrowding accelerates the spread of tuberculosis and other respiratory infections. The absence of sanitation and lack of a safe water supply provides fertile ground for outbreaks of water- and food-borne enteric diseases, and the larger the population exposed, the greater the risk and subsequent extent of infection. In the mid-1980s, it was estimated that 17 million people (of which 10.5 million were infants under the age of

Figure 18.4
Deaths from infectious diseases and parasites in developing countries in the mid-1980s

based on data from (10)

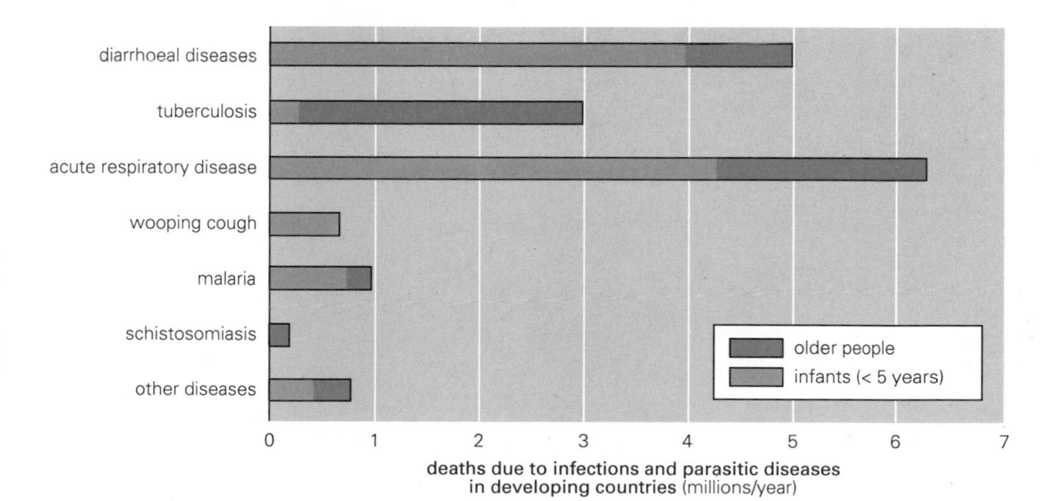

five) in the developing countries died every year of infectious and parasitic diseases (Figure 18.4), as compared to about half a million in developed countries (10).

Although cholera has subsided in Asia, it has made its way into the Americas, resulting in a dramatic increase in the number of cases reported to the WHO (about 250 000 cases in 1991). In Africa the total number of cholera cases has been almost stable for the past two decades. Occasional local outbreaks of cholera have occurred in some countries, mainly as a result of the contamination of drinking water and food.

Malaria continues to be one of the most serious public health and environmental problems in large parts of the developing world. The disease is endemic in 102 countries, placing over half the world's population at risk. Since 1980 there has been a general decrease in malaria in Africa, South-East Asia and the Western Pacific, but a gradual increase in the Americas (Figure 18.5). In 1988, 8 million cases of malaria were reported to WHO, but it is believed that the overall number of cases in the world is in the order of 100 million (12). Of those reported in 1988, 39 per cent were in Africa and 32 per cent in South-East Asia (Figure 18.6). It is believed that 43 per cent of the world population lives in areas where malaria occurs, and some 445 million of them live in areas where no specific transmission control measures are undertaken and where the prevalence of the disease remains virtually unchanged (12).

Schistosomiasis remains a major health threat in some 76 developing countries. The largest number of cases occur in Brazil, Central Africa, China, Kampuchea, Egypt and the Philippines. It is

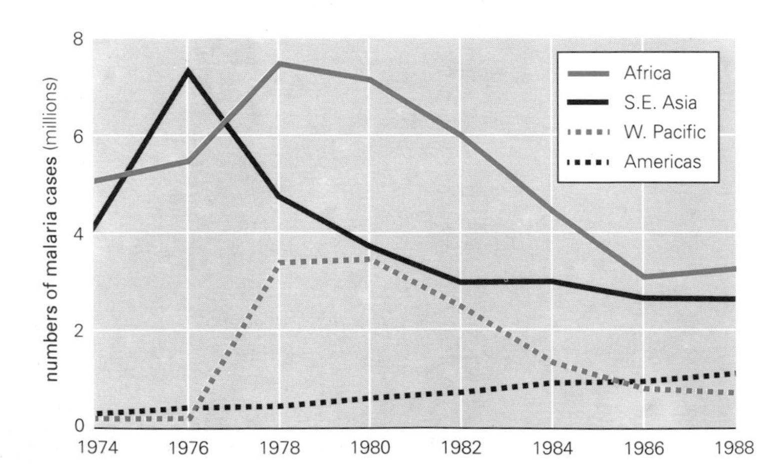

Figure 18.5
Trends in malaria incidence in most affected regions

based on data from (13, 14)

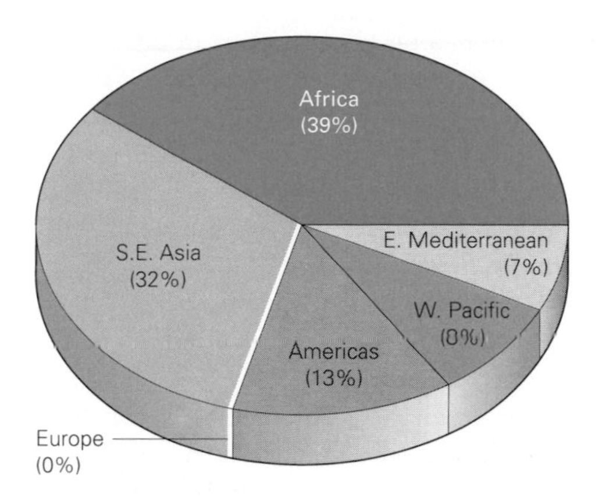

Figure 18.6
Distribution of malaria, by region

based on data from (14)

estimated that some 200 million people are infected and another 600 million are at risk of infection. The creation of man-made lakes, fish-ponds and irrigation schemes has contributed to an increase in the incidence of the disease. Following the construction of the Diama dam on the Senegal River in 1986, it was found that intestinal schistosomiasis had increased since early 1988; by 1989, 71.5 per cent of samples were positive (15, 16).

The first reported cases of HIV infection and AIDS (acquired immunodeficiency syndrome) occurred in the past decade. AIDS kills people of all ages, but is a growing threat to newborn children and infants. At least 1.5 million women worldwide—of whom about one million live in Africa—are infected with HIV. Babies born to such women have a 25–40 per cent chance of being infected before or during birth. These children are almost certain to die by the age of five (17). It is estimated that worldwide some 5–10 million people are infected with the AIDS virus and about 400 000 have full-blown AIDS (14). Estimates are that by the end of 1991, more than a million cases of AIDS will have occurred worldwide, and that by the year 2000 the cumulative total could exceed five million (14).

Chemical pollution and health

People are exposed to a number of hazardous chemicals at home and at work. A vast amount of scientific information is available on the short-term effects of exposure to high levels of hazardous chemicals—but little is known about what happens to individuals exposed to very low concentrations of chemicals over longer periods of 20–30 years. However, the consequences of such exposure can be measured in the population at large in terms of physiological change, disease and death. Genetic mutations—the production of new, mostly detrimental, hereditary traits—can also have chemical causes, and they are permanent. Cancer and birth defects are among the other

At least 1.5 million women worldwide—of whom about one million live in Africa—are infected with HIV. Babies born to such women have a 25–40 per cent chance of being infected before or during birth. These children are almost certain to die by the age of five.

health hazards that can result from long-term exposure to toxic substances. Birth defects occur in 2–3 per cent of all births. Of these, 25 per cent have underlying genetic causes, while 5–10 per cent result from the influence of four main known external agents: chemicals, drugs, radiation and viruses. The remaining 65–70 per cent arise from unknown causes, but may result from the interaction of a number of environmental agents with genetic factors (18, 19, 20).

The effects of exposure to a chemical pollutant depend on the length and severity of exposure and the chemical involved. There are two main types of exposure. The first includes exposure to abnormally high levels of pollutants—as in accidental chemical releases (see Chapter 9), exposure at work and severe air pollution episodes. The second type of exposure is to general ambient pollutants.

In the first case, the effects are evident and include death and an increase in morbidity. The accidental release of methyl isocyanate in the Bhopal accident (see Chapter 9) led to both deaths and a high rate of morbidity. Exposure of workers to high concentrations of chemicals has also led to various occupational diseases. Lead poisoning, pneumoconiosis (a lung disease caused by dust inhalation), pesticide poisoning and various cancers can result from such exposure. WHO has estimated that the number of unintentional cases of acute poisoning due to pesticide exposure was half a million in 1972 and increased to one million in 1985, because of greater use of pesticides. About 60–70 per cent of these cases are due to occupational exposure. Some 20 000 deaths a year occur as a result of pesticide poisoning (21).

Many traditional occupational diseases are declining in developed countries, as a result of the strict enforcement of protection measures. But occupational diseases are on the increase in several developing countries, because of a lack or non-enforcement of regulatory measures to protect workers, and also because of a lack of awareness and cooperation among workers. There is also mounting concern about the increase of occupational diseases in small-scale industries, including repair workshops, especially among children who constitute a large proportion of the work force (see Chapter 12).

The effects of air pollution episodes, such as the London smog of 1952, are well documented; children and the elderly (in particular those with respiratory and circulatory problems) were most affected.

Assessing the health impacts of exposure to chemical pollutants in the general environment is difficult because an individual is generally exposed to several pollutants at the same time. Total exposure includes inhalation, ingestion and skin absorption of the pollutants from air, water, food or soil. In many cases the effect of an individual pollutant can be increased or decreased through interactions with other pollutants; the adverse health effects of sulphur dioxide are known to increase in the presence of particulate matter; tobacco smoking increases the incidence of cancer due to exposure to indoor radon (see Chapter 1).

Various attempts have been made in the past two decades to estimate the health impacts of total human exposure. Models have been used to calculate the environmental distribution, transformation and fate of chemical pollutants; human exposure via different routes; and the toxicological and dynamic properties of chemical substances in humans (22, 23).In 1984, WHO/UNEP set up the Human Exposure Assessment Locations programme (HEALs), as part of GEMS, in order to monitor total human exposure to pollutants. The results should enable countries to assess the combined risks from air, food and water pollutants, and to take appropriate action to safeguard human health.

Cause and effect have been established for several pollutants. The health effects of carbon monoxide, tropospheric ozone, sulphur oxides combined with particulates, and lead in ambient air are well documented (see Chapter 1). Epidemiological research over the past two decades has established that indoor air pollution could increase the incidence of cancer due to radon exposure and tobacco smoke. In rural areas of developing countries it could increase respiratory diseases and cancer due to exposure to emissions from biomass fuel. The increase in nitrates in groundwater has become a cause for concern in several countries as nitrates constitute a health risk, especially for infants. WHO, UNEP and ILO have been working together since the early 1970s to establish health criteria for various pollutants (see Chapter 10).

There is now widespread agreement that roughly 85 per cent of all cancers are caused by environmental factors, such as ionizing radiation, carcinogenic chemicals in air, food or water, smoking, alcohol and drugs, including chemotherapeutic agents. The rest, presumably, have a hereditary basis or else arise from spontaneous metabolic events. Although the percentage of deaths from cancer is higher in developed than in developing countries (Figure 18.3), the pattern of cancer incidence in both groups of countries is similar. However, variations in the incidence of different types of

Occupational diseases are on the increase in several developing countries ... There is also mounting concern about the increase of occupational diseases in small-scale industries, including repair workshops, especially among children who constitute a large proportion of the work force.

cancer are encountered (Figure 18.7). Tobacco smoking (including passive smoking) is the most important cause of lung cancer. In spite of this well-established fact, the global use of tobacco has grown by nearly 75 per cent over the past two decades and smoking has increased markedly among young people.

Responses

The different responses outlined in previous chapters contribute, directly and indirectly, to the improvement of human health and to reducing the health risks associated with exposure to different pollutants. Results have demonstrated that prevention is better than cure. Though the original goals of the IDWSSD were not met by 1990, the 1980s did provide hundreds of millions of people with safe drinking water and sanitation facilities (see Chapter 5) and this has largely contributed to health improvements in the areas provided with such facilities. Figure18.8 illustrates how improved water supplies and sanitation facilities can reduce the rate of illness related to diarrhoea. Further preventive measures include actions taken to reduce air emissions (see Chapter 1) and to protect the ozone layer (see Chapter 2). However, there is still a long way to go before the health risks of environmental pollution and deterioration are significantly reduced. Much research is required to clarify the causes and effects of total human exposure to establish practical guidelines to protect human health.

Figure 18.7 Incidence of new cancer cases in the early 1980s in developed and developing countries

based on data from (24)

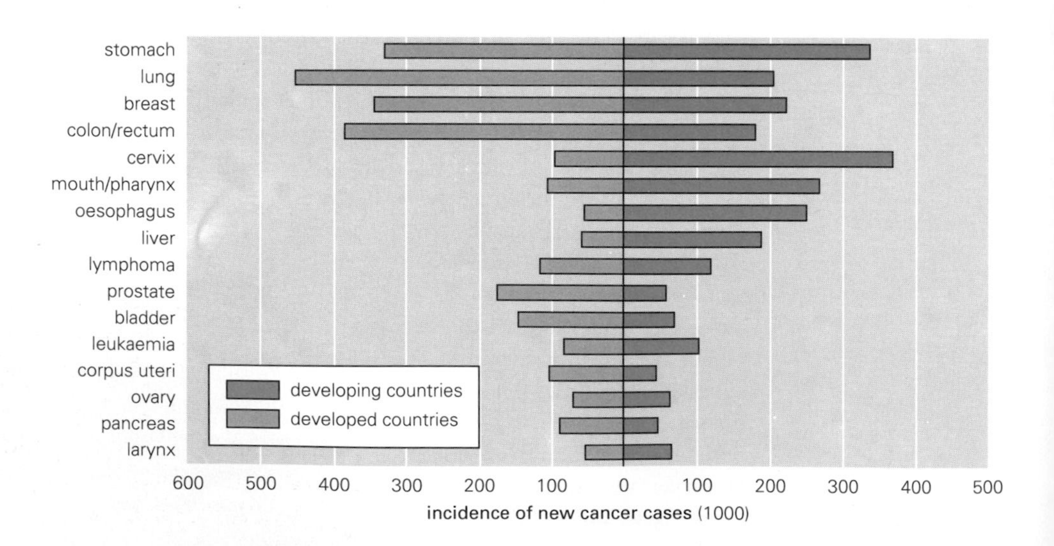

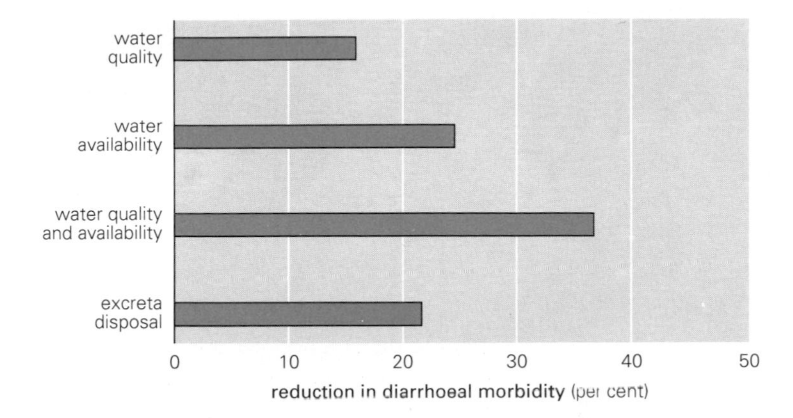

Figure 18.8
The effect of improved water supply and sanitation facilities on diarrhoeal morbidity

based on data from (25)

There is still much to be done to reduce the incidence of communicable diseases in developing countries, though several communicable diseases were brought under control in the past two decades. Smallpox was eradicated. The incidence of onchoceriasis (river blindness) has been reduced sharply in West Africa. An increase in the use of oral rehydration therapy (ORT) reduced the mortality of children under the age of five due to diarrhoeal diseases. In 1985, about 18 per cent of children with diarrhoea were treated with ORT; by 1989 the percentage of these children reached 25 per cent (26), saving the lives of some one million children each year (27). The six vaccine-preventable diseases of childhood—poliomyelitis, tetanus, measles, diphtheria, pertussis and tuberculosis—have declined through increased immunization. In the 1970s, these diseases killed about five million children a year, but in the 1980s the figure dropped to about three million a year and continues to reduce through the expanded programme of immunization.

In the 1970s, the six preventable diseases of childhood—poliomyelitis, tetanus, measles, diphtheria, pertussis and tuberculosis—killed about five million children a year, but in the 1980s the figure dropped to about three million a year and continues to reduce.

Chapter 19

Peace, security and the environment

Violence is a prehistoric solution to disputes which time and culture have endowed with endless sophistications, but have otherwise left unchanged. The scientific progress of recent centuries has merely enabled us to kill more people, more quickly and effectively than our medieval ancestors or our fellow primates and other mammals. Only recently has it been realized that war and preparations for war are inimical to development: they squander scarce resources and erode the international confidence that is essential to promote development, conserve our scarce resources and protect the environment at regional and global levels.

A world at war

Over the past two decades, the world spent about US$17 trillion (at 1988 prices and exchange rates) on military activity. In other words, global military expenditure averaged US$850 billion per year—US$2.33 billion a day, US$97 million an hour or US$1.6 million a minute. In current terms, annual global military expenditure reached more than US$1000 billion in 1990 (1, 2, 3). Military expenditure has consistently increased since 1970, though there has been a slight deceleration since the mid-1980s (Figure 19.1). Military spending in industrialized countries, and as a share of global GNP, has decreased slightly but it has increased in most developing countries (Figure 19.2). On a regional basis, Latin America devotes the smallest share of its GNP—about 1.5 per cent—to military spending. The Middle East and North Africa, on the other hand, spend the highest share of GNP—about 12.6 per cent—on military activities (4, 5, 6). All in all,

Figure 19.1 World military expenditure at 1988 prices (below) and distribution of world military expenditure (below right)

based on data from (1, 3)

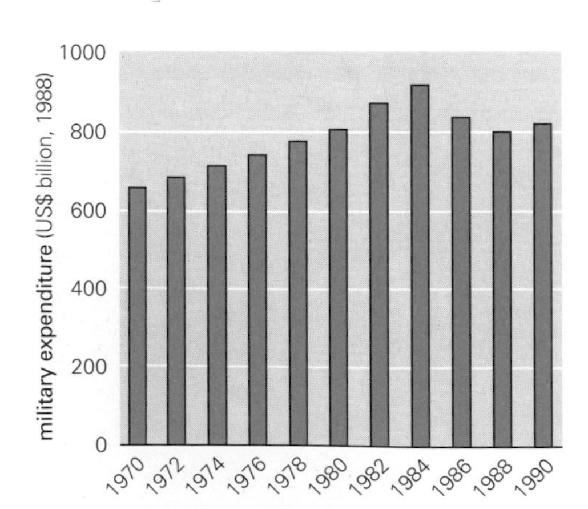

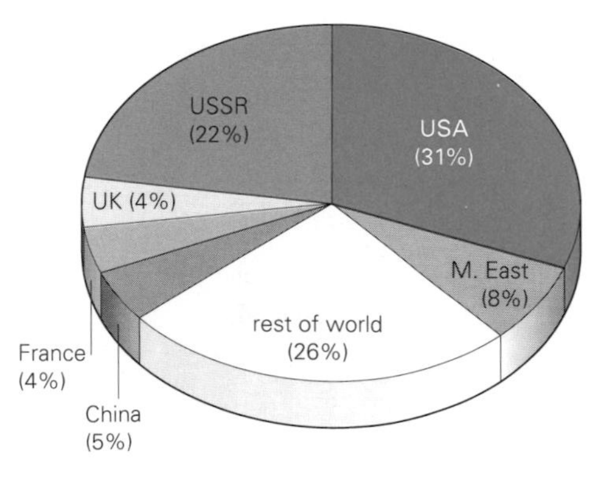

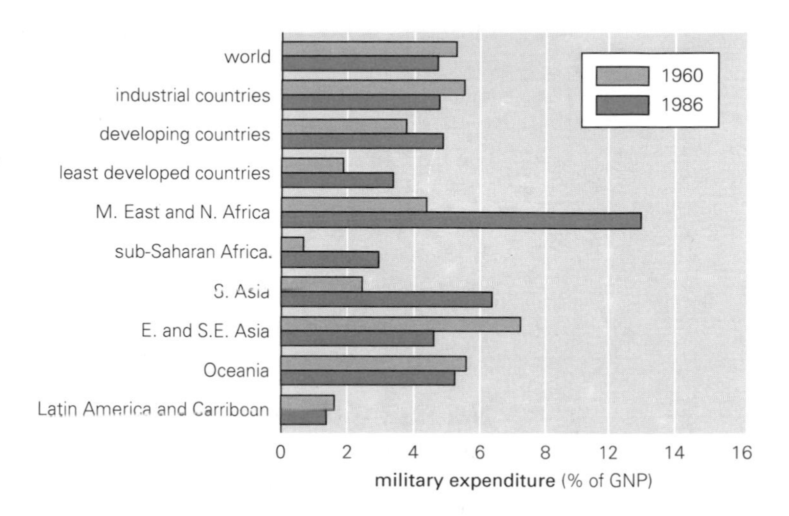

Figure 19.2
Military expenditure as a percentage of GNP (1960 and 1986)

based on data from (5, 6)

the world's military spending dwarfs any spending on development.

The increase in world militarization has been accompanied by a dramatic growth in the arms trade. Over the past two decades, cumulative global arms sales have reached US$410 billion—about US$20 billion a year (Figure 19.3). It has been estimated that about 50 per cent of all arms imports into developing countries have been financed by export credits (7). The cost of such military credit amounts to 30 per cent of all debt inflow to developing countries.

Militarization has also diverted considerable resources away from development activity. The military employs some 60–80 million people worldwide (7), including about 3 million scientists and engineers.

Figure 19.3
Regional distribution of arms exporters and importers (below) and global arms sales (below right)

based on data from (3)

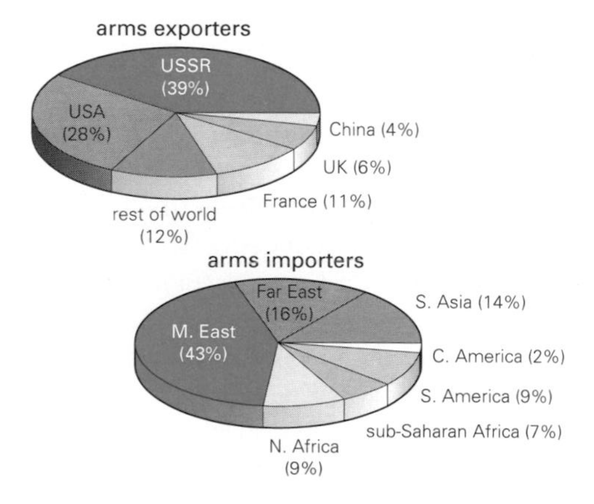

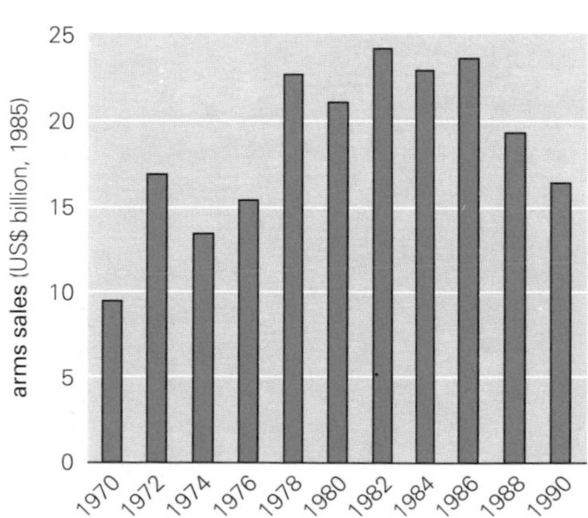

Large areas of land are set aside for military training and weapons testing. In several countries prime land is used for the construction of military installations and service buildings, without any consideration of better use of such lands for national socio-economic development. The military also uses vast quantities of energy and mineral resources. It has been estimated that the global military demand for aluminium, copper, nickel and platinum has been greater than the total demand for these minerals for all purposes in Africa, Asia and Latin America combined. About 6 per cent of the total world oil consumption is for military purposes—nearly half the total oil consumption of all the developing countries.

War and the environment

Almost all wars have had one basic strategy: destruction of the enemy's life-support systems so their armies and peoples can be defeated. Carpet bombing of towns and their infrastructures was widely used in World War II; extensive bombing, and chemical and mechanical methods of destroying forests and crops, were widely used in the Indochina War of 1961-75 to drive the fighters and their farmer supporters out of their villages and hiding places. With advances in military technology, a whole series of guided weapons has emerged that can hit targets more precisely without causing full-scale collateral damage. The effectiveness of these high-tech weapons was demonstrated in the war over Kuwait in 1991. Damage to the built environment caused by all conventional wars can be repaired, but two types of war destruction are much more difficult to repair: damage to the natural environment and damage to the social fabric of the affected population.

The extensive use of chemical warfare (herbicides) in the Second Indochina War has shown the catastrophic damage that can occur to the environment as a result of war. Millions of litres of different herbicides were sprayed over an area of about 1.7 million ha in Indochina in the period 1961–71 (8), resulting in large-scale devastation of crops and forests. This has led to widespread soil erosion, decimation of terrestrial wildlife, loss of freshwater fish and declines in coastal marine fisheries; since then recovery of the affected ecosystems has been slow. The impact on humans has varied from neuro-intoxications to increased incidence of hepatitis, liver cancer, spontaneous abortions and congenital malformations.

The war over Kuwait in 1991 resulted in a huge oil spill and extensive, continuous fires in oil wells. The oil spilled from loading

The extensive use of chemical warfare in the Second Indochina War has shown the catastrophic damage that can occur to the environment as a result of war. Millions of litres of herbicides were sprayed over Indochina in the period 1961–71, resulting in large-scale devastation of crops and forests.

terminals, sunken and leaking vessels in the northern part of the Gulf has been estimated at 4–8 million barrels. The spill damaged coastal areas in some countries and also affected wildlife and aquatic life (9). The fires started in 613 Kuwaiti oil wells caused the burning of about 4 to 8 million barrels per day and resulted in massive clouds of smoke and gaseous emissions that spread over a large area in the northern Gulf (9, 10, 11). Measurements carried out showed that about 1-2 million tonnes of carbon dioxide were emitted each day, together with varying amounts of sulphur and nitrogen oxides, carbon monoxide and organic compounds. Near the Kuwait border, emissions averaged about 100 000 particles per cubic centimetre. Most of the smoke mass was transported at an altitude of 2-3 km for distances of up to 2000 km, mainly towards the east and the south-east. The most direct impact of the smoke was to reduce incoming solar radiation, which lowered the surface temperature in some parts of the northern Gulf. Direct health effects included some respiratory symptoms in sensitive groups, but a detailed assessment of the effects remains to be made (9). By November 1991, the fires had been brought under control and all wells were capped.

Millions of unexploded bombs, booby traps, land and sea mines, and other types of munitions are left behind after the cessation of military hostilities (see box). Scant information is generally available on the number and location of such remnants of war, which makes their clearance difficult and risky. The remnants of war have endangered people, livestock and wildlife, and hindered the development of vast areas of land (12).

The remnants of war

- In Poland, 14 894 000 land mines and 73 563 000 bombs, shells and grenades have been recovered since 1945.
- In Finland, more than 6000 bombs, 805 000 shells, 66 000 mines and 370 000 high explosive munitions have been cleared since the end of World War II.
- In Indochina, about 2 million bombs, 23 million artillery shells and tens of millions of other high explosive munitions were left unexploded after the war.
- In Egypt, following the 1973 Arab-Israel War, about 8500 unexploded items were removed from the Suez Canal and more than 700 000 land mines were cleared from the terrain near the Canal. Thousands of land mines and unexploded shells are still scattered around the Gulf of Suez and in the Sinai peninsula.

Source (39)

Wars and conflicts have generated millions of displaced people, or refugees. Their exact number is not known, partly because of the lack of an internationally-accepted definition of a refugee (13). Estimates indicate that the number of refugees increased from about 3 million in 1970 to about 15 million in 1990 (Figure 19.4). These refugees have not only suffered economic losses, but the whole social fabric of their lives has been disrupted. In most cases these refugees live in camps in border areas where living conditions are poor and disturbances are common. In some cases these people cannot be rehabilitated to their original homes, and they continue to live in misery for decades.

Nuclear weapons have added an entirely new dimension to warfare. The two atom bombs dropped on Hiroshima and Nagasaki in 1945 had a yield of 12.5 kilotonnes and 22 kilotonnes of TNT respectively. Their devastating effects have been well documented. Nuclear weapons developed subsequently have dramatically increased destructive powers, not of kilotonnes but megatonnes. The number of nuclear warheads in the world has been estimated at between 37 000 and 50 000, with a total explosive power of 11 000– 20 000 megatonnes—equivalent to 846 000–1 540 000 Hiroshima bombs.

Despite the widespread condemnation of nuclear weapons, their production and testing have continued. The total number of nuclear tests from 1945 to 1990 was 1818, of which 489 were in the atmosphere and 1329 were underground (Figure 19.5). In the 1980s, several studies were carried out to predict the impacts of a large-scale

Figure 19.4
Estimated numbers of refugees by region, 1970–90

based on data by UNHCR

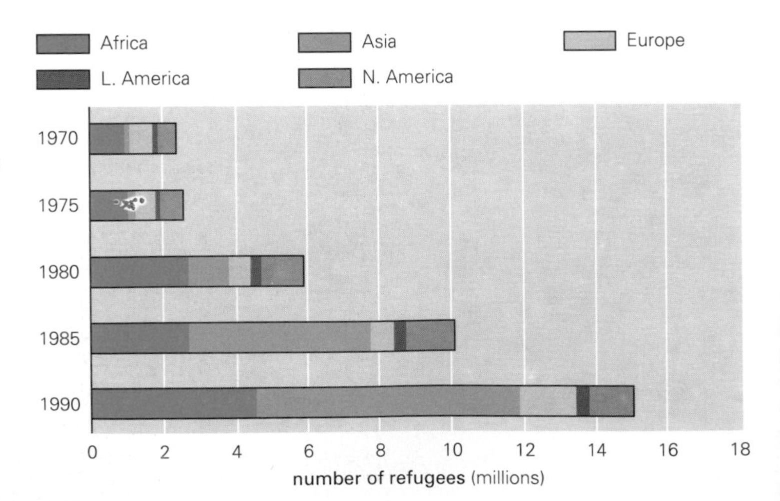

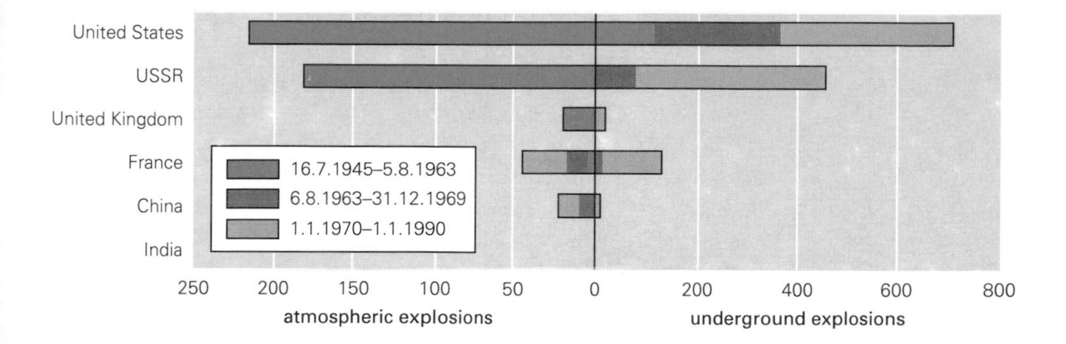

Figure 19.5
Numbers of atmospheric and underground nuclear explosions; the Test Ban Treaty was signed on 5 August 1963

based on data from (3)

nuclear war (14–26). In spite of a number of uncertainties, different nuclear war scenarios give estimates of about 30-50 per cent of the total human population as immediate casualties of a nuclear war. The 50-70 per cent who might survive the direct effects of a large-scale nuclear war would be affected by a 'nuclear winter'. In the aftermath of a large nuclear war, darkened skies would cover large areas of the earth for perhaps weeks or several months, as the sunlight was blocked by large, thick clouds of smoke from widespread fires (23, 24). Temperatures would drop below freezing and rainfall could be affected in many regions. Such climatic changes would affect agriculture and major ecosystems, such as forests, grasslands and marine ecosystems, with far-reaching impacts on food production and distribution systems.

In the 1970s there were speculations about the possibility of causing economic or other damage to an enemy population through environmental disturbance (27–30). Environmental warfare could, in principle, involve damage caused by manipulating influences from celestial bodies or outer space, or influences in the atmosphere, on the land, in the oceans, or in the biota. Another concern has been the possibility of using biological weapons—living organisms, usually pathogenic micro-organisms—for hostile purposes; such concern has deepened with advances in biotechnology and genetic engineering. The effectiveness of existing biological agents can now be enhanced, and new, more potentially effective ones could be created.

In the past two decades it has become evident that military means are no longer adequate to provide tangible security benefits—the security of nations depends as much on economic well-being, social justice, and ecological stability.

Evolving concepts of security

Several studies of the relationship between the arms race and development (31–34) have stressed the fact that each competes for the world's finite resources. In the past two decades it has become evident that military means are no longer adequate to provide tangible security benefits—the security of nations depends as much on economic well-being, social justice, and ecological stability. Environmental degradation imperils the most fundamental aspects of national security by undermining the natural support systems on which all human activity depends. Because pollution and environmental degradation respect no man-made borders, they jeopardize not only the security of the country in which they occur, but also that of others, near and far. Spurred by a stream of new scientific evidence, attention is now shifting to those aspects of environmental degradation that have an all-encompassing, global effect from which no nation can insulate itself. Even though their full impacts may be felt only years or decades from today, the depletion of the ozone layer (see Chapter 2) and the trend to global warming (see Chapter 3) can no longer be considered as hypothetical threats.

This thinking has lead to the evolution of new concepts of security. Expressions such as 'balance of power', 'deterrence', 'peaceful coexistence', 'collective security', and 'common security' have been introduced (35) to emphasize that security comprises not only

Major multilateral arms control agreements, 1970–90

- Treaty on the prohibition of the emplacement of nuclear weapons and other weapons of mass destruction on the sea bed and on the ocean floor and soil. Signed 1971, entered into force 1972.
- BW Convention on the prohibition of development, production and stockpiling of biological and toxin weapons, and on their destruction. Signed 1972, entered into force 1975.
- Protocols I and II to the 1949 Geneva Convention relating to the protection of victims of armed conflicts. Signed 1977, entered into force 1989.
- Enmod Convention on the prohibition of military or other hostile use of environmental modification techniques. Signed 1977, entered into force 1978.
- Inhumane Weapons Convention on the prohibition or restrictions on the use of certain conventional weapons which may be deemed to be excessively injurious or to have indiscriminate effects. Signed 1981, entered into force 1983.
- South Pacific nuclear-free zone treaty (Treaty of Rarotonga). Signed 1985, entered into force 1986.

Source (3)

military, but also political, economic, social, human rights, humanitarian and ecological aspects.

Environmental stress is both a cause and effect of political tension and military conflict. Nations have often fought to assert or resist control over raw materials, energy supplies, land, river basins, sea passages and other key environmental resources (36). Such conflicts are likely to increase as these resources become scarcer and competition for them increases. Disputes have also arisen over the use or pollution of shared water resources, acidic precipitation, marine pollution, siltation of downstream river beds, increased floods and management of groundwater resources.

Responses

Many conventions, treaties and agreements have been adopted to limit and prevent the devastating effects of war (see box). But mounting military expenditure implies a general lack of conviction in keeping constant the size of forces and arsenals, let alone in reducing them. A further conflict exists between the increasing demand for resources for development and the increasing allocation of such resources for military purposes. A major breakthrough in disarmament would release vast financial, technological and human resources for more productive uses in both developed and developing countries in an international political climate of reduced tension.

The rechannelling of resources from the military to the civilian economy has been referred to as the conversion process. This conversion has political, economic and technical dimensions (34). Unilateral measures to curtail military spending and initiate a conversion process can be taken by any state, but in a real global political sense, disarmament has to be started by the major powers, on the basis of mutual, verifiable agreements to reduce armaments and eliminate particular military capabilities.

Conversion is more than a theory. In 1985, China decided to utilize part of its military industrial capacity to manufacture civilian goods. Civilian production now accounts for 20 per cent of output from China's military factories, and that share is projected to reach 50 per cent by 2000 (37, 38). Conversion produces more jobs, helps to meet

Spending US$1 billion on guided missile production creates about 9000 jobs. Spending the same amount on educational services creates 63 000 jobs; and on air, water and solid waste pollution control, 16 500 jobs. A US$40 billion conversion programme could bring a net gain of more than 650 000 jobs.

Contradictions and trade-offs between military, social and environmental priorities

- The United Nations Environment Programme, the organization responsible for safeguarding the global environment, spent US$450 million over the past ten years—less than five hours of global military spending.
- Total annual official development assistance extended to developing countries is US$35 billion—equivalent to 15 days of global military spending.
- Between six and seven hours of world military spending—US$700 million—could be used to eradicate malaria, the disease that claims the lives of one million children every year.
- One and a half days of global military spending—US$3.10 billion—equals the annual cost of protecting land unaffected by desertification and reclaiming those areas moderately affected.
- Three days of global military spending—US$7 billion—could fund the Tropical Forestry Action Plan for five years.
- One Apache helicopter—US$12 million—costs the same as installing 80 000 hand pumps to give Third World villages access to safe water.
- One Patriot missile system—US$123 million, without missiles—costs the equivalent of 5000 low-cost housing units to free 5000 families from life in the slums.
- One day spent on the 1991 war over Kuwait—US$1.5 billion—could have funded a five-year global child immunization programme against six deadly diseases, thereby preventing the death of one million children a year.

Source (37, 38, 40)

growing socio-economic needs, and is of vital importance in conserving resources and in environmental protection. In the United States, spending US$1 billion on guided missile production creates about 9000 jobs. Spending the same amount on educational services creates 63 000 jobs; and on air, water and solid waste pollution control, 16 500 jobs. A US$40 billion conversion programme could bring a net gain of more than 650 000 jobs (38). The trade-offs between military and social and environmental priorities can, indeed, be far-reaching (see box).

In the face of transnational environmental problems, national responses are likely to prove fruitless without international cooperation. It is true that all states have the responsibility to ensure that activities within their jurisdiction or control do not cause damage to the environment of other countries (Principle 21 of the Stockholm Declaration). But it is also true that environmental security is critically dependent on pragmatic internationalism. The Convention on Transboundary Air Pollution (see Chapter 1), the Montreal Protocol to protect the ozone layer (see Chapter 2) and the conventions on biodiversity and climate change under negotiation are all examples of international efforts to foster global environmental security. The regional conventions for the protection of the marine environment, the regional seas programmes (see Chapter 4), and the cooperative programmes for the environmentally-sound management of inland waters (Chapter 5) are all steps being taken in the same direction.

The world community still needs urgently to review the status of different international treaties dealing with the environment in cases of war. In particular, the Hague Conventions II of 1899 and IV of 1907, the Geneva Protocol of 1925 on Chemical and Bacteriological Warfare, the World Cultural and Natural Heritage Convention of 1972, the Berne Protocol I of 1977 and the Environmental Modification Convention of 1977 should all be reviewed and strengthened. The General Assembly of the United Nations is dealing with this issue as this report goes to press.

All states have the responsibility to ensure that activities within their jurisdiction or control do not cause damage to the environment of other countries. It is also true that environmental security is critically dependent on pragmatic internationalism.

Part IV

Perceptions, Attitudes and Responses

Chapter 20

Perceptions and attitudes

People have always cared about the environment, though their perceptions of environmental issues and their attitudes have evolved over the centuries. In the earlier part of this century environmentalism was essentially synonymous with wildlife conservation and considered to be the domain of a prescient, and often privileged, few. Since the 1960s environmentalism has become a movement with widespread popular support and an extensive range of interests. The United Nations Conference on the Human Environment convened in Stockholm in 1972 was a turning point in the history of environmental awareness. Growing public pressure, backed by scientific findings in the late 1960s and early 1970s on the impacts of pollutants and environmental degradation, stimulated the necessary political will. The debate that took place in the early 1970s, based essentially on air and water pollution in the North, brought home the fact that environmental degradation is caused not only by industrialization but by poverty and lack of development. The environmental movement has since become concerned with all aspects of the natural environment: land, water, minerals, living organisms, life processes, the atmosphere, climate, polar ice-caps, remote ocean deeps, and even outer space. Furthermore, the environmental movement has expanded its examination of the natural environment in isolation to include its interrelationship with human well-being and with the status of international economic cooperation covering issues of debt, commodity prices, structural adjustments, subsidies and so on.

Environmentalism has not only grown in the past two decades but has also altered its complexion to suit the times. Modifications to social cost-benefit analysis, the onset of environmental impact assessment and environmental auditing, risk analysis, public inquiries, new legislative measures at national and international level, and the activities of non-governmental groups have all helped to give policies and actions a more environmental tenor.

Recent years have seen the development of another phase of the environmental movement. This is characterized by the concern evinced by some important, complex, and widespread problems, and by the organization taking place on a national and international scale around these issues. Examples are acid rain, the disposal of hazardous wastes, global warming, loss of biodiversity, depletion of the ozone layer, marine pollution, deforestation and the interaction between peace, security and environment. Effective action on these issues requires a wide range of skills: considerable academic knowledge on the part of those actively involved; an ability to organize activities in the often widely-separated areas in which issues surface; political skill to deal with the governments, industries, special interest groups and individuals who play major roles in such issues; an ability to communicate with, and through, the media; and, perhaps most important of all, a long-term concern and willingness to face the 'big' issues. All these are characteristics of emerging environmentalist professionalism (1).

Scientific groups and non-governmental organizations (NGOs) have played a major role in the environmental movement from its start. Environmental groups have a wide range of interests. Small ones are organized to fight local problems, often environmental disruption—immediate or potential—from pollution or some inappropriate form of development. Others deal with a specific issue on a national scale. Some national evironmental groups are primarily concerned with the use of the environment and who should benefit from it; others have been described as 'sustainable development' or 'appropriate technology' groups. Over the past decades, an increasing number of international environmental NGOs have emerged—including powerful bodies such as Friends of the Earth, Greenpeace and the World Wide Fund for Nature (WWF). A unique link between the non-governmental and governmental sector has been provided since 1948 by The World Conservation Union (IUCN) which links some 55 States, 100 government agencies and 450 NGOs. The range of conservation, development and humanitarian NGOs, and of industry groups concerned with the environment, has expanded steadily during the 1970s and the 1980s, and contacts between NGOs and governments have also been strengthened.

Through environmental groups, therefore, individuals are increasingly able to influence national and international policies. But today the environmental movement is so diverse that the question arises as to whether it is really proper to give it a single name. Environmental organizations and their members often differ in the extent of their concern over particular environmental issues, and in their values, attitudes, goals, objectives and choice of strategies and tactics. Yet what is shared by all these organizations is a concern about socio-environmental relationships. UNEP has responded to such diversification through its 'outreach' policy of opening up a dialogue with industry, parliamentary groups, relief

Today the environmental movement is so diverse that the question arises as to whether it is really proper to give it a single name ... Yet what is shared by all these organizations is a concern about socio-environmental relationships.

organizations, women's groups, youth organizations, religious bodies and other groups that are receptive to the environmental message. This message is being accepted and adopted by more and more people and sections of society in both developed and developing countries. The environment is becoming an increasingly popular issue. Perceptions and attitudes are changing, and the changes are proving remarkably widespread and robust.

Public perceptions and attitudes towards environmental issues are conditioned by cultural, traditional, socio-economic and political factors. Since the 1960s, perceptions and attitudes towards environmental issues have changed considerably. Students of modern environmentalism (2–7) have identified three types of perceptions. In the first, environmentalism is characterized by an emphasis on the need for strong environmental legislation and technological solutions (such as recycling), and by the idea that reforms can be produced by idealism, determination, goodwill and the efforts of individuals, local groups and committees. In the second, environmentalism is characterized by the realization that in modern society things are not that simple. There has consequently been a growth in the formation of pressure groups that aim to influence the decision-making process. The third type of environmentalism is characterized by the development of a critique of technology and energy use in present society, and calls for the development of alternative or 'soft' technologies and increased self-reliance.

One way to get an indication of what people think about environmental issues is through a referendum or poll. Despite various limitations linked to the size, structure and characteristics of the sample of the population surveyed, public opinion polls still provide

Figure 20.1
The results of a multinational poll of the public and leadership on environmental issues

based on data from (14)

Key

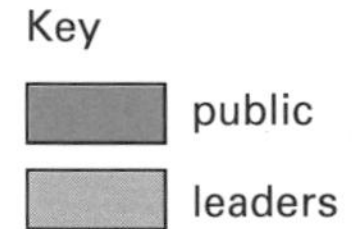

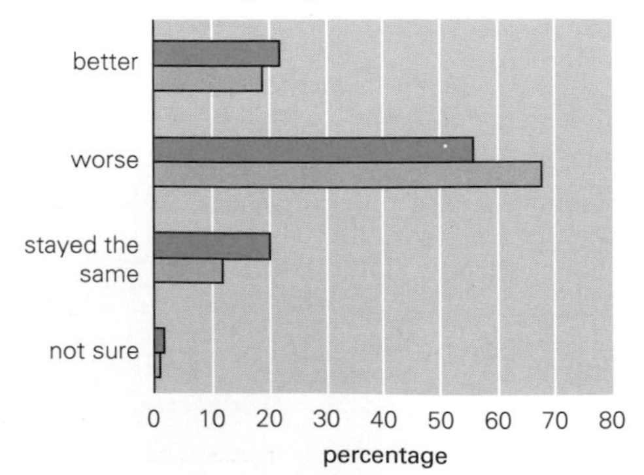

What are the major environmental problems?

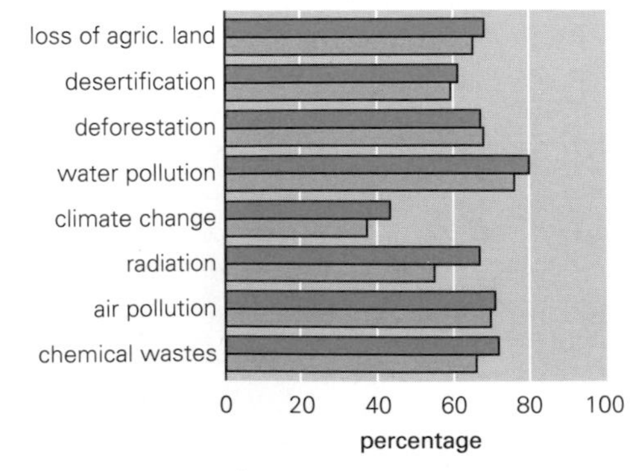

A multinational survey of environmental issues showed remarkable agreement between the public and leadership about the state of the environment.

the most useful measure of changing public attitudes. While public opinion polls carried out in the late 1960s and early 1970s concentrated mainly on local environmental issues, those conducted more recently have often included national, regional and global environmental issues, as well as issues related to socio-economics, politics, development and quality of life (8).

High levels of public concern and consciousness about environmental issues have been recorded in all polls (8-13). A multinational survey of public and leadership perception of environmental issues covering 14 countries—Argentina, China, Hungary, India, Jamaica, Japan, Kenya, Mexico, Nigeria, Norway, Saudi Arabia, Senegal, West Germany and Zimbabwe (14)—showed a remarkable agreement between the public and leadership about the state of the environment, the problems considered to be 'major', the need for international cooperation to deal with environmental problems, and even the division in opinion about the willingness to pay more to protect the environment (Figure 20.1).

On the other hand, a recent study (15) revealed marked differences in opinion between the public and experts on certain environmental issues. While the public expressed extreme concern about the risks associated with nuclear power, radioactive and hazardous waste, and chemical plant accidents, experts ranked these issues as medium to low risk. Conversely, issues of greatest concern to experts, such as pesticides, indoor air pollution, worker exposure to chemicals, and global warming, were regarded as medium- to low-risk issues by the public.

Should environmental protection be carried out alone or in cooperation with other countries?

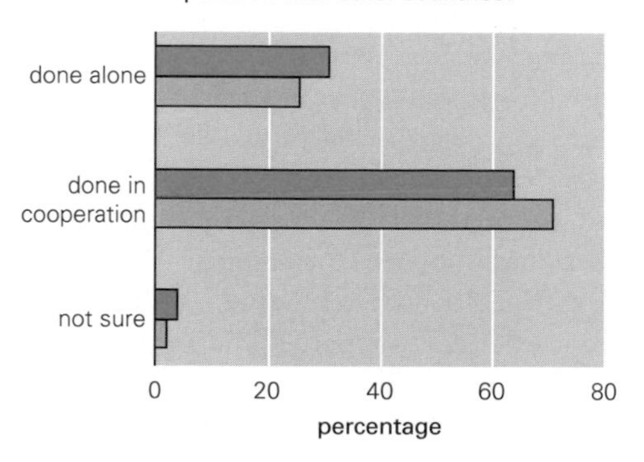

Are you willing to pay higher taxes to protect the environment?

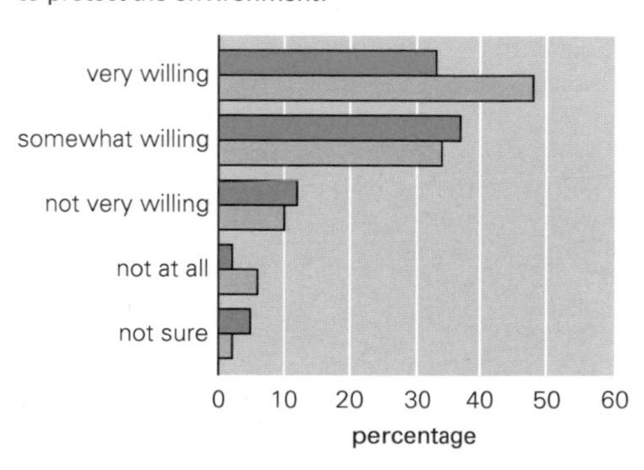

There are several reasons for this divergence of opinion. The first is that the public does not have all the information necessary to form an accurate opinion. The second is the difference in perception of hazards. Public concern is aroused when a significant hazardous environmental accident occurs, and public perceptions are greatly influenced by media coverage of such an accident. This is natural, because public perception of a hazard is heavily influenced by its potential severity and very little by the frequency of its occurrence. Perhaps irrationally, people often overestimate the frequency of occurrence and seriousness of the risk posed by dramatic, sensational and well-publicized causes of death, and underestimate the risks posed by more familiar, accepted causes that claim lives one by one (see also Chapter 9).

Such divergences have often frustrated decision makers and have led some experts to argue that the public's apparent pursuit of a 'zero-risk society' threatens national, political and economic stability. There is, indeed, no such thing as 'zero risk'. However sophisticated and advanced a technology may be, it cannot be foolproof, and there is no safeguard against human error. Experts and decision makers should, therefore, understand how people think about and respond to risk rather than devote their attention only to statistical estimates of it. Without such understanding, well-intended decisions and policies may become ineffective. In any case, the public should ultimately decide what risks it will accept. If people are encouraged to become fully involved in planning and decision making—through participation—the public and the experts will learn from each other, society as a whole will benefit and more durable policies will emerge.

The media have been instrumental in increasing public awareness of many environmental issues, but the media have been generally reactive rather than innovative. Coverage rose and fell in response to 'trigger events', either disasters or threatened disasters. Events such as smog episodes in London in 1952 and New York in 1963, the Seveso accident in 1976, the *Amoco Cadiz* accident in 1978, the Three Mile Island nuclear power accident in 1979, the Bhopal tragedy in 1984, the Chernobyl accident in 1986 and the *Exxon Valdez* disaster in 1989 received extensive coverage by the press, radio and television, partly because they have inherent public appeal. This natural predisposition towards the dramatic ensures that information provided by the media about risks is frequently inadequate. When environmental risk is reported, the emphasis is usually on its more alarming features. The stark language of news—that is, the words and pictures used to convey information—also leaves room for interpretation, especially in the absence of background information (8, 16-21).

Improving the supply of environmental information to the media, together with accurate reporting of such information to the public, is critical for the management of environmental problems. Successful environmental communication is not to be measured by public acceptance of the solutions formulated by decision makers; it is achieved when the best solutions are chosen by a well-informed public. One of the most important roles of the media is to expand the audience for debate on a particular issue. This frequently leads to a redefining and broadening of the scope of the problem, and often creates new issues and more controversy. These new issues and the expanded audience for policy discussions have frequently frustrated decision makers and made them feel that the resolution of problems has become more difficult. But, at the same time, the raising of these issues has resulted in new thinking to include the new dimensions that emerged from the debate within a wider audience. This usually leads to better, more enduring policies.

Responses

Evolution in public perceptions of environmental issues, increased public awareness and the activities of different national and international NGOs gave impetus to many actions in the past two decades to protect the environment. All the responses outlined in the previous chapters and those given in Chapter 21 are generally the result of a public outcry for a better environment and living conditions. In the early 1970s, some predicted that environmentalism would pass through an 'issue attention cycle', during which it would leap into prominence, remain there for a short time and then gradually fade from public attention as economic recession, developing country debt and regional conflict grew (22). Yet, environmentalism has grown in every respect and it is here to stay. The 1990s is witnessing not only a more vigorous interest but also an important transformation in thinking. An increasing proportion of people in many countries now accept the need for development strategies that enable people to live off 'nature's interest', rather than

Environmentalism has grown in every respect and it is here to stay.

'nature's capital' (23). More people are accepting the principle of intergenerational responsibility and equity, on the basis that future generations should not inherit less environmental capital than the present generation inherited.

One manifestation of increased public concern for the environment, particularly in developed countries, is an increase in the demand for information on which to base the choice of products 'friendly' to the environment. This rise in 'green consumerism' (24–26) has led to the incorporation of environmental considerations into the policies of several national and international consumer pressure groups. One example of their success is the return to refillable containers (instead of tins) for soft and other drinks in some European countries, including Denmark. Another example is the increased use of recycled paper for packaging and other purposes.

The growth of the environmental movement has had a profound influence on industry. Whereas in the 1960s and 1970s, industry tended to regard environmental concerns as a peripheral nuisance, to be evaded where possible, in the 1980s many companies have themselves become active in developing environmental policies. Corporate managers are beginning to see that improving the environment is the smart way of conducting business. Based on the premise that profit-making opportunities of the 1990s will be in manufacturing and marketing 'environmentally-sound' products and services, initiatives such as developing cleaner production processes, offering products that generate less waste, devising safer pest control strategies and cleaning up past damage are fast becoming top-priority investment areas (27, 28). Recycling of waste (Chapter 10) and increasing the efficiency of water, energy and raw material use in manufacturing processes (Chapter 12) are examples of the response of industry to the environmental movement. Another important example is the cooperation of industry in phasing out chlorofluorocarbons and other compounds that have been implicated in depleting the ozone layer (Chapter 2).

Chapter 21

Responses

People and governments have always responded to environmental deterioration. The ancient Chinese, for example, appointed inspectors to ensure that cultivated land was not degraded through the use of unsuitable farming methods. Ancient Greek and Roman scholars wrote about soil husbandry and land management. Plato described in *The Laws* what could be considered as the earliest known enunciation of the 'polluter pays principle':

> 'Water is easily polluted by the use of any kind of drug. It therefore needs the protection of a law, as follows: whoever purposely contaminates water shall be obliged in addition to paying an indemnity, to purify the spring or receptacle of the water, using whatever method of purification is prescribed.'
> (*The Laws*, Book VIII, p. 845)

The first smoke abatement law was passed in England in 1273 (1). Cities passed many ordinances against refuse dumping in streets and canals. The environmental students of the 19th century expressed concern about the impacts of human transformation of the landscape, and early geographers and geologists attempted to describe the changing face of the earth in its entirety. The destruction of natural areas stimulated the formation and growth of conservation measures; early conservationists worked in defence of natural reserves, ancient buildings and different habitats. It was not until the early 1960s, however, that the growing environmental movement (Chapter 20) placed increasing pressures on governments to respond to different contemporary and emerging environmental issues. The responses outlined in the preceding chapters cover the wide range of activities undertaken in the past two decades to curb environmental pollution at national, regional and global levels, and conserve and manage natural resources in a more rational way. The following is a an analysis of the different categories of responses set in motion.

Science and technology

In the past two decades scientific research has contributed a great deal to our understanding of the processes that control and affect environmental systems. Much progress has been made in improving analytical methods and instruments to determine and monitor trace amounts of inorganic and organic pollutants; in defining the processes of transformation and interaction of pollutants, and the fate of pollutants emitted into various media; and in establishing the

As a result of increased understanding of complex environmental systems, and of advances in computer technology, model performance has improved greatly in the past 20 years.

effects of such pollutants on material and biota. Impressive insights have been gained into the biogeochemical cycling of elements essential for life such as carbon, nitrogen, oxygen, phosphorus and sulphur. And we now understand the mechanisms that could lead to ozone depletion and global warming better than we did two decades ago.

Many of these scientific advances have been achieved through national and international scientific research programmes. The activities of UNEP, the Scientific Committee on Problems of the Environment (SCOPE), the joint WMO/ICSU Global Atmospheric Research Programme (GARP), the World Climate Programme, the UNESCO Man and the Biosphere (MAB) programme, the WHO/ILO/UNEP International Programme on Chemical Safety, the International Geosphere-Biosphere Programme (IGBP), the CGIAR, the International Federation of Institutes for Advanced Study (IFIAS), the International Institute for Applied Systems Analysis (IIASA), IUCN, IIED, WRI and several United Nations bodies have contributed a great deal to our understanding of contemporary and future environmental problems.

Mathematical models have long been used to predict geophysical, as well as some ecological, processes. As a result of increased understanding of complex environmental systems, and of advances in computer technology, model performance has improved greatly in the past 20 years and there is now an increased degree of public acceptance of model-derived 'futures'. The most recent models relate to ozone depletion, climate change, acid rain, nuclear winter and impacts of environmental change on the biosphere. Models developed in the past two decades dealt with the interrelationships between resources, population growth and environment, and included World I and II models, which formed the basis of *Limits to Growth* published by the Club of Rome, the Leontief model published by the United Nations, the OECD Inter-Futures model, the Latin American Model published by the Bariloche Foundation, the Global 2000 model published by the US Council on Environmental Quality, the IIASA models related to energy supply and demand, and several other regional and global models.

The past two decades have also seen the development of environmental impact assessment, cost-benefit analysis, risk analysis and management, natural resources and environmental accounting, technology assessment, environmental audits, geographical information systems and several other tools that have contributed to a better understanding of environmental processes and, to a marked degree, to better policies to deal with environmental problems. For example, the use of environmental cost-benefit analyses has brought about considerable improvements in regulatory measures in the United States. The adoption of more stringent limits on lead in fuels brought a net benefit in health and welfare of an estimated US$6.7 billion (2).

Advances have also been made in technologies to protect the environment. For example, more efficient air pollution control equipment (such as electrostatic precipitators and flue-gas desulphurization equipment) has been developed. Improved technologies have been introduced to treat both municipal and industrial wastewater, manage solid wastes, increase efficiency of energy and water use, and beneficially use several types of wastes. A number of 'cleaner' technologies have been developed. The steps taken by the world community to phase out chlorofluorocarbons, which threaten the ozone layer, have been accompanied by a notable response from research laboratories and chemical companies to develop environmentally-sound alternatives. Major advances have been made in developing simple technologies, particularly for use in rural areas of the developing countries. The efficiency of conversion of dung into biogas and fertilizer has been improved by changes in the design of digesters, more efficient wood-burning stoves have been developed, various types of hand-pumps for water supply have been introduced and simple latrines have been designed to improve sanitation. Several technologies have also been developed to harness renewable sources of energy, especially solar and wind power, for crop drying, water pumping and other purposes.

Education and training

Parallel to these advances has been a remarkable increase in environmental education—both formal and informal—in almost all countries. The environmental subjects that used to be embodied in courses such as chemistry, biology and botany now constitute separate environmental courses in many schools and universities. Specific undergraduate and post-graduate courses on the environment are now offered by many universities around the world. Training programmes on environmental issues have been held at universities and research centres. The UNESCO/UNEP International Environmental Education Programme (IEEP) has been associated with worldwide efforts to incorporate an environmental dimension into education since 1975. The widespread use of informal information channels (especially television and the press) has contributed significantly to increased public awareness of environmental issues.

Institutional measures

The task of designing and implementing environmental protection programmes rests with national governments. In the early 1970s, only a few—mainly developed—countries had government departments that were concerned with aspects of environmental management. Sweden established a National Environmental Protection Board in 1969, while the United States established its Council on Environmental Quality and its Environmental Protection Agency in 1970 under a National Environment Policy Act. The United Kingdom established a Royal Commission on Environmental Pollution and a Department of the Environment in 1970, and Canada established its Department of Environment at about the same time. In 1971, Japan established its Environmental Agency, and France established a Ministry for the Environment and the Protection of Nature. It was not until after the Stockholm Conference that departments of environment and cross-sectoral coordinating machinery for environmental affairs were established in many countries. At present, nearly all countries have environmental machineries of some kind. Some countries have established ministries for environment and/or natural resources; others have established environmental protection agencies and/or departments, either as independent bodies or affiliated to particular ministries. The responsibilities of these environmental bodies vary from one country to another. In general, their function is to protect the national environment through the enactment of legislation, the establishment of emission levels and the creation of monitoring programmes to identify the most serious problems and measure the success of programmes to control such problems. The success of national environmental bodies has varied considerably. In many countries, especially developing ones, inter-departmental conflicts have arisen that have weakened the operational ability and limited the coordinating role of environmental bodies.

At present, nearly all countries have environmental machineries of some kind ... In general, their function is to protect the national environment through the enactment of legislation.

Although some UN bodies such as FAO, UNESCO and WHO dealt—within their mandates—with different issues related to the environment before 1972, it was not until the Stockholm Conference and the establishment of the United Nations Environment Programme (UNEP) that a major driving force was created to encourage UN bodies to incorporate environmental concerns into their activities and create units or departments to deal with environmental issues. UNEP, with its catalytic and coordinating role, has contributed a great deal to that driving force. The major United Nations conferences that followed the Stockholm Conference explored in depth issues related to food, freshwater, human settlements, desertification, renewable sources of energy and other topics, and led to the broadening of the mandates of UN bodies and/or the creation of additional intergovernmental and secretariat bodies within the UN system to deal with contemporary and emerging issues. These developments also encouraged a broadening and strengthening of the mandates and activities of global institutions such as IUCN, ICSU (especially its Scientific Committee on Problems of the Environment), WWF and others. Global NGOs, including the International Institute for Environment and Development (IIED), the World Resources Institute, Greenpeace and Friends of the Earth were established at this time and are now instrumental in providing independent advice on environmental and resource issues.

At the regional level, the Economic and Social Commissions of the United Nations established special units to deal with environmental issues. In South-East Asia, joint action by the South Pacific Commission, the South Pacific Bureau for Economic Cooperation, ESCAP and UNEP led to the establishment of the South Pacific Regional Environmental Programme in 1982. The African Ministerial Conference, convened in Cairo in 1985, led to the adoption of a regional programme of action and the establishment of a secretariat to follow up its implementation. The Conference of the Arab Ministers responsible for Environment, convened in Tunis in 1986, adopted a regional programme of action and established a ministerial council to follow up its implementation. Other regional inter-governmental bodies, including OECD and CMEA, have established units and various committees to deal with regional environmental issues.

Another important development in the past two decades was an increasingly widespread recognition of the importance of incorporating environmental considerations into development policies and assistance. Accordingly, UNEP and nine development assistance agencies signed the Declaration of Environmental Policies and Procedures Relating to Economic Development, in 1980. They pledged to create systematic environmental assessment and evaluation procedures for all development activities and to support projects enhancing the environment and natural resource base of developing nations. The Committee of International Development Institutions on the Environment (CIDIE) was established to review the implementation of the Declaration at regular intervals.

The World Bank has established a department to deal with environmental issues pertaining to its activities, and almost all regional development banks have incorporated environmental impact assessment into their development-supported projects. Recently, a multilateral fund involving UNEP, UNDP and the World Bank was established to help developing countries meet the costs of complying with the revised Montreal Protocol and to provide for any necessary transfer of technology (Chapter 2). The Global Environmental Facility also became operational recently. This is a joint World Bank/ UNDP/UNEP venture with a fund of about US$1.3 billion to address priority global environmental issues.

Regulatory measures

Although several countries formulated laws to improve the quality of their environment many decades ago, most of these have been amended or clarified in recent years. In some cases, changes were required because problems were found to be more serious than had originally been thought. In other cases, adjustments made environmental protection programmes more effective. Much national environmental law has been concerned with regulating activities that may pose an environmental hazard and limiting discharges and emissions to the environment. Another dimension of environmental law relates to the procedures to be followed before development projects are implemented or before products are marketed. These include requirements for environmental impact assessments, and screening and approval of products including pharmaceuticals and pesticides. Another area of recent legislation concerns public access to information, and the right of the public to sue companies and other

The Global Environmental Facility ... became operational recently. This is a joint World Bank/UNDP/UNEP venture with a fund of about US$1.3 billion to address priority global environmental issues.

bodies that cause environmental damage or put the environment at risk through their activities.

The emergence of an increasing body of environmental law and regulation has been paralleled by changes in the way in which legal instruments are interpreted and enforced. Conflicts often arise, and it has been difficult to implement national environmental regulations, especially in developing countries. Sometimes, national environmental bodies do not have sufficient information to assess the extent to which polluters do not comply with existing rules. In many developing countries, environmental regulations emulate those in developed countries, and therefore cannot be implemented because of considerable differences in environmental and socio-economic conditions. For example, several developing nations have failed to enforce regulations on vehicle emissions similar to those implemented in the United States and some European countries.

The recognition that environmental pollution is not restricted by national boundaries, but can cross frontiers and cause regional and global problems, has prompted the formulation of regional and international conventions. Before 1972, there were 58 international treaties and other agreements in the field of environment; between 1972 and 1991, 94 such agreements were adopted regionally and globally (see UNEP, 1991 Register of International Treaties and other Agreements in the Field of Environment). These international agreements deal with issues including marine pollution, and protection and conservation of natural resources, and with future issues (such as early warning in case of nuclear accidents and protection of the ozone layer). The increasing concern of the world community about global warming and its potential impacts on different ecosystems has prompted the start of negotiations to draft a global climate convention. Similarly, negotiations are under way to elaborate an international convention on the conservation and rational use of biological diversity (see Chapters 1-10). In spite of the large number of legal instruments in the field of environment, the effectiveness of environmental legislation leaves much to be desired. Lack of compliance with existing laws and weakness of enforcement procedures are two major concerns. Although international environmental agreements, as a rule, incorporate obligations to report on implementation, mechanisms for verifying implementation and resolving environmental conflicts have yet to be adequately institutionalized.

Economic measures

Over the past 20 years an increasing number of countries have recognized that economic instruments can be an effective means of improving the environment and maintaining high environmental quality. Several guiding principles have evolved since 1970. The first was the 'polluter pays principle', or PPP, which essentially states that the costs of pollution should not be externalized. An industry or municipality should itself bear the costs, without subsidy, of actions needed to meet environmental standards and avoid environmental damage. As a consequence, market prices should reflect the full cost of environmental damage arising from pollution or, more appropriately, the cost of preventing such damage. Similarly, the 'user pays principle', or UPP, requires that prices reflect the full social cost of use or depletion of a resource.

In OECD countries, several economic instruments are used. Charges, including effluent charges, user charges, product charges and administrative charges, are used to discourage polluting activities and/or to provide financial assistance to achieve reductions in pollution. Subsidies, in the form of grants, soft loans and tax allowances may be used to encourage less-polluting behaviour. Deposit-refund schemes (for example, on beverage containers) encourage re-use or more environmentally-friendly disposal of waste. Market creation arrangements, such as trading arrangements, encourage more efficient and cost-effective use of emission permits. Financial enforcement incentives, such as non-compliance fees and performance bonds, provide an financial inducement to comply with existing environmental regulations (3). By 1988, 153 different economic instruments were said to be in use in OECD countries. Of these, 81 involved charges, 41 were subsidies and 31 were other measures (3). For example, France has an effluent charge related to air pollution; Finland and Sweden have a carbon tax on fossil fuel use; Australia, Belgium, The Netherlands and the United States levy effluent charges on wastes; Denmark, Finland, Germany, The Netherlands, New Zealand, Sweden, Switzerland and the United Kingdom impose different levels of taxation on leaded and unleaded petrol; and Germany, The Netherlands, Sweden and Japan use taxation as an instrument to promote low-pollution vehicles (3).

France has an effluent charge related to air pollution; Finland and Sweden have a carbon tax on fossil fuel use; Australia, Belgium, The Netherlands and the United States levy effluent charges on wastes.

Economic measures have also been implemented in many developing countries. Some of the first were charges for the collection of domestic rubbish, deposit-refund schemes (especially for beverage and other containers) and several types of fine for illegal dumping of waste (such as that created by the construction sector). In the past 20 years an increasing number of countries have reduced or removed subsidies on agricultural chemicals such as pesticides and this has led to more efficient use of these compounds and/or the increasing adoption of integrated pest management techniques (Chapter 11). However, the application of economic measures to curb pollution by industry and vehicles has been more difficult.

Part V

Challenges and Priorities for Action

Chapter 22

Challenges and priorities for action

Delays have dangerous ends
(W. Shakespeare; Henry VI, Part I, 1589/90)

Ten years ago the UNEP publication *The World Environment 1972–1982* concluded with the observation:

> At the Stockholm conference it was generally assumed that the world's system of national governments, regional groupings and international agencies had the power to take effective action ... By the early 1980s there was less confidence in the capacity of national and international managerial systems to apply known principles and techniques, or in the effectiveness with which international debates lead to action ... Restoration of confidence and consensus in these areas may be the greatest challenge for those seeking to improve the world environment in the 1980s.
>
> UNEP 1982 (1)

It is disturbing that the same statement is still valid a decade later and that many of the concerns identified in the earlier report remain the same. There are still serious gaps in our understanding of the environment, in our ability to estimate the cost of repairing the damage we have done to it and in our assessment of the cost of failing to take rapid action to halt its degradation. Twenty years after the Stockholm Conference, it is still not possible to describe the state of the world environment comprehensively or to say with confidence that governments of the world have the knowledge and political will to deal with the global problems that we already know exist.

The most significant concerns are a lack of many basic prerequisites for good environmental management and informed decision making, in particular:

- The world environment database is still of variable quality, and there is a shortage of data from developing countries. As a result, only 'best estimates', not comprehensive data, on major environmental problems can be compiled.
- Great recent scientific advances, including developments in remote sensing and in the technical ability to monitor the world environment, have not been uniformly applied—mainly because many countries lack equipment and trained personnel.
- No general agreement has been reached on standards for a decent environment or on the socio-economic indicators of a healthy relationship between people and environment.

- Comprehensive assessments of the environmental situation and of the earth's carrying capacity are, in consequence, still difficult.

Despite these problems there has been progress in a number of areas over the past decade. Scientific assessment of stratospheric ozone depletion and understanding of the processes involved have progressed very rapidly and have been matched by international and national actions to redress the situation. Strong scientific consensus is now emerging on climate change and loss of biodiversity, their causes and the need for a collective response. Some progress has been made in coping with hazardous wastes and toxic chemicals. There are also a greater number of more reliable overall environmental assessments backed up by improved data.

The first section of this volume outlined the ten major environmental issues of the past two decades, their impacts on people and the possible responses to them. Major trends over the past 20 years have included:

- Levels of urban air pollution have been lowered in most cities in the developed world, but there has been a marked deterioration in developing countries. Further work and much greater international cooperation are needed to deal with transboundary and global atmospheric pollution.
- Rapid advances in the scientific understanding of stratospheric ozone depletion and its causes indicate that further action to protect the ozone layer will be required if governments are to avert significant health and economic effects.
- Increased understanding of the causes and possible effects of climate change, despite uncertainties in complex climate modelling, indicate a range of global strategies to be adopted urgently to counter its effects.
- Access to fresh water and the quality of available water supplies are key factors in development, particularly in arid or semi-arid areas. Urgent action is therefore needed to improve both knowledge and management of freshwater resources. Cooperative management of freshwater basins should be established to avoid potential conflicts.
- Urgent action is still needed to deal globally with land-based sources of marine

... it is still not possible to describe the state of the world environment comprehensively or to say with confidence that governments of the world have the knowledge and political will to deal with the global problems that we already know exist.

pollution and the unsustainable use of marine resources, and to rehabilitate degraded areas—despite the progress already made through a number of regional sea action plans to halt further degradation of coastal zones, seas and oceans.

- Desertification and degradation of arid lands are grave and growing problems, and their socio-economic and physical causes must be addressed urgently. A realistic programme of corrective action and rehabilitation in land subject to desertification must begin.
- Deforestation, wetland destruction and habitat loss are threatening the stability of local and regional environments and wasting valuable resources. Regional, subregional and national actions, within agreed global targets, are urgently needed to halt and reverse these problems in all regions.
- Loss of biological diversity, expressed in the rapid extinction of species and reduction of genetic variability, is an unnecessary waste of irreplaceable resources needed for sustainable development. Urgent action is needed to save, study and use rationally the world's biological riches.
- Where human activity has increased the range and scale of environmental hazards to which people are exposed, action is needed to reduce the risk of disasters, particularly from human causes, and to improve our responses to unavoidable disasters.
- The generation and disposal of hazardous wastes, and the production of toxic chemicals, pose significant threats to human well-being. International action to improve our knowledge and control of hazardous wastes and toxic chemicals is urgently needed.

These issues arise from human actions, and are of concern mainly because their effects impinge on human well-being—either directly or by undermining the life-support systems of the environment.

Environment and development

Development is a multidimensional concept, encompassing economic, cultural, political and social aspects of human society. The preceding chapters have illustrated how, through the development process, human beings interact with and affect the natural environment, and how the state of the environment determines the path of development. The world community is confronted by a closed cycle: economic problems cause or aggravate environmental despoliation; this, in turn, makes economic and structural reform difficult to

If the world continues to accept disappearing tree cover, land degradation, expansion of deserts, loss of plant and animal species, air and water pollution, and the changing chemistry of the atmosphere, it will have to accept economic decline and social disintegration.

[T]he kind of environmental [p]roblems that are of [i]mportance in developing [c]ountries are those that can [b]e overcome by the process of [d]evelopment itself.
[T]he Founex Report [1]971 (2)

[T]he protection and [i]mprovement of the human [e]nvironment is a major [i]ssue which affects the well-[b]eing of peoples and [e]conomic development [t]hroughout the world.
[T]he Stockholm [D]eclaration 1972 (3)

[T]he problem today is not [p]rimarily one of absolute [p]hysical shortage but of [e]conomic and social [m]aldistribution and misuse.
[T]he Cocoyoc [D]eclaration 1974 (4)

[I]t is essential that the [m]utually dependent [r]elationships between [d]evelopment and [e]nvironment be fully [a]nd explicitly taken [i]nto account.
[U]NEP 1980 (5)

[H]ealth, nutrition and [g]eneral well-being depend [u]pon the integrity and [p]roductivity of the [e]nvironment and resources.
[I]nternational [D]evelopment Strategy [f]or the Third UN [D]evelopment Decade, [U]N 1980 (6)

achieve. If the world continues to accept disappearing tree cover, land degradation, the expansion of deserts, the loss of plant and animal species, air and water pollution, and the changing chemistry of the atmosphere, it will also have to accept economic decline and social disintegration. In a world where progress depends on a complex set of national and international economic ties, such disintegration would bring insecurity and human suffering on a scale without precedent.

Until the early 1970s it was familiar for debate about environmental policy to be couched in terms of economic growth versus environmental protection. The basic idea was that one could have economic growth—measured by rising real per capita income—or one could have improved environmental quality; any mix of the two involved a trade-off where more environmental quality meant less economic growth, or vice versa. However, in the 1971 Founex Seminar on Development and Environment, the 1972 United Nations Stockholm Conference on the Human Environment, the 1974 Cocoyoc Symposium on Patterns of Resource Use, Environment and Development Strategies, organized by UNEP and UNCTAD in Mexico, and in other fora and studies, the links between environment and development became clarified (see box left). Since then, discussion has tended to shift from growth versus the environment to the potential complementarity of growth and environment.

The 1970s saw the emergence of a major revision in development thinking that has presented a fundamental challenge to the conventional consensus on economic development. New expressions have been introduced such as 'alternative patterns of development and lifestyle', 'eco-development', 'environmentally sound development', 'development without destruction', and 'sustainable development' to convey the same message that environment and development are closely interdependent, and are in fact mutually supportive (see boxes on pages 247 and 248).

Over the past two decades, the concept of sustainable development has been increasingly stressed. Although there are many definitions of sustainable development (10), it is generally understood to involve the key elements identified in the report

Eco-development ... is a style of development which stresses specific solutions for the particular problems in each eco-region taking into account ecological and cultural contexts as well as present and long-term needs.
UNEP Report to Governing Council 1974 (7)

Development without destruction—the maximization of the production of food without destroying the ecological basis to sustain production ...
M. K. Tolba, Statement to World Food Conference 1974 (8)

Environmental management implies sustainable development
UNEP 1975 (9)

Special attention from now on must be placed on ... adjusting lifestyles to a more rational use of resources with particular emphasis on the present and future resource and environmental needs of the ... developing countries.
UNEP 1980 (5)

of the World Commission on Environment and Development (11) and in *Environmental Perspective to the Year 2000 and Beyond* (12) (see box right).

Central to the concept of sustainable development is the requirement that current practices should not diminish the possibility of maintaining and improving future living standards. In other words, economic systems should be managed to maintain and improve the environmental resource base so that future generations will be able to live equally well or better. Sustainable development does not require the preservation of the current stock of natural resources nor any particular mix of human, physical and natural assets. Nor does it place artificial limits on economic growth—provided that such growth is economically and environmentally sustainable.

Sustainable development raises the possibility of a new type of fairness and equality rarely considered previously—that of inter-generational equity. In the past it was commonly assumed that the next generation would take its chances on a planet very similar to the one inhabited by the current generation—perhaps with new technology to make life safer, healthier and easier. This assumption is no longer justifiable. The present generation is the first to have the power to alter planetary ecosystems radically, to present its offspring with a planet very different from the one it inherited from its own forebears—different in atmosphere, soils, water systems, plants and animals.

But inter-generational equity is a difficult goal—unborn generations cannot be present to make their concerns known. The need for environmentally sound and sustainable development means that this generation must accept responsibility for future generations. Realizing this goal may be the foremost challenge facing policy makers in the closing years of the 20th century and beyond.

The integration of environmental management and economic and social development was raised at the Stockholm Conference, but it is still a major arena of debate. There have been many developments in the past two decades which promise major changes in the way societies think about managing the future relationship of human activity and the natural environment. But most of these advances have yet to be institutionalized into government and development agency policies and planning systems. Few countries take adequate

To defend and improve the environment for present and future generations has become an imperative goal of mankind—a goal to be pursued together with, and in harmony with, the established and fundamental goals of peace and of worldwide economic and social development.

The Stockholm Declaration

Sustainable development ... meets the needs of the present without compromising the ability of future generations to meet theirs ... on the basis of prudent management of available global resources and environmental capacities and the rehabilitation of the environment previously subjected to degradation and misuse ... Although it is important to tackle immediate environmental problems, anticipatory and preventive policies are the most effective and economical in achieving environmentally sound development.
UNEP 1987 (12)

Critical objectives for environment and development policies that follow from the concept of sustainable development include:
- *reviving growth*
- *changing the quality of growth*
- *meeting essential needs for jobs, food, energy, water and sanitation*
- *ensuring a sustainable level of population*
- *conserving and enhancing the resource base*
- *reorienting technology and managing risk*
- *merging environment and economics in decision making*

WCED 1987 (11)

Growth must be revived in developing countries ... where the links between economic growth, the alleviation of poverty and environmental conditions operate most directly.
WCED 1987 (11)

The process of economic development must be more soundly based on the stock of capital that sustains it.
WCED 1987 (11)

account of environmental considerations when making policy or planning development. Few allocate or regulate use of their living resources to ensure that they are environmentally appropriate and sustainable. Many lack the financial or technical resources, the political will, or adequate legislative, institutional, and public support to tackle environmental problems. The result has been that, at the level of project planning and design, unwanted environmental impacts have arisen from inadequate attention paid to environmental consequences, and from a lack of the knowledge and information necessary to predict them. Other causes have included ignorance of cost-effective preventive or mitigation measures, and failure to consider alternative project designs or locations (13).

Environment and economics

Classical economic theories and practices have treated nature as an infinite supply of physical resources (such as raw materials, energy, water, soil and air) to be used for human benefit, and as an infinite sink for the by-products of the development and consumption of these benefits, in the form of various types of pollution and ecological degradation. Hence, the economy became disembodied from nature, in theory and in practice. The dominance of this approach began to weaken in the late 1960s, when pollution became a major concern in the industrialized nations. It was soon realized that the self-regeneration of natural resources is a slow and complicated process; if some natural resources are over-exploited, the stock will fall rapidly, leading ultimately to the complete destruction of the resource. It was also realized that air and water have limited assimilative and carrying capacities, and that pollution control measures must be instituted to safeguard the environment and the quality of human life.

It is therefore important, if sustainable development is to be achieved, to evaluate the environmental costs and benefits of any development process. But such evaluation is not easy. Some of the environmental effects of development can be easily identified and evaluated quantitatively; others cannot. Nevertheless, an economic analysis of the environmental effects of alternative development processes, partial though it must necessarily be, is important because it creates awareness of the fact that natural resources ought not to be treated as free goods. Environmental costs arise either through the damage done as a consequence of resource exploitation or through the effort expended to redress the damage.

In the past two decades several studies have attempted to estimate the economic costs of damage caused by environmental pollution. For example, the annual damage caused by air, water and noise pollution in The Netherlands was estimated at US$0.6–1.1 billion in 1986 (about 0.5 to 0.9 per cent of the GNP). In Germany, the damage from the same sources of pollution has been estimated at about US$34 billion per year in 1983/85, or about 6 per cent of annual GNP (14). The economic cost of pollution damage in developed countries varies between 3 and 5 per cent of GNP. However, this costing of the damage due to irrational use of natural resources and/or pollution is far from complete. Environmental damage is often selective, and unequally distributed in time and space and among societies. Many of the physical, biological and socio-economic consequences of large development projects are inadequately known; some can be quantified while others cannot. Examples of the latter are when landscape or historic monuments are threatened with irreversible change. Even if all the consequences could be enumerated and their likelihood assessed, placing a price tag on them would pose further difficulties. Consider, for example, the problems of placing a value on a human life. The traditional economic approach has been to equate the value of a life with the value of a person's expected future earnings. Many problems with this index are readily apparent. For one, it undervalues those in society who are underpaid and places no value at all on people who are not earning. In addition, it ignores the interpersonal effects of a death which may make the loss suffered much greater than any measurable financial loss.

The cost of pollution abatement and control in the developed countries has been estimated at 0.8 to 1.5 per cent of annual GDP (15). For developing countries, the figure is much lower and varies markedly from one country to another.

Pollution abatement studies focus essentially on the direct costs of dealing with pollution problems such as air and water pollution, and management of waste. In most cases they do not include the cost of environmental deterioration, loss of natural resources nor of the impacts of all this on economic development and on human health and well-being. Such studies, therefore, generally show the cost of action to protect the environment and its natural resources but not the

For society as a whole, environment is a part of its real wealth and cannot be treated as a free resource.
The Founex Report 1971 (2)

Incomplete accounting occurs ... especially in the case of resources that are not capitalised in enterprise or national accounts: air, water, and soil.
Our Common Future 1987 (11)

cost of inaction. The important point is that the costs of environmental policies are in fact an investment for the future. The costs are generally more than compensated for by the benefits accrued from reducing the damage and from conservation of resources. For example, it has been estimated that the net benefits from air and water pollution control in the United States would amount to about US$26 billion per year (14). In the developing countries, the construction of drinking water and sanitation facilities could reduce the incidence of infectious diseases by 50–60 per cent or even more (16). Such an improvement in human health would lead not only to an increase in productivity and time on the job (both of which contribute to increased GNP), but also to a smaller expenditure on goods and services delivered by the medical sector, most of which are imported.

In the past two decades, some attempts have been made to adjust national income accounts to register both the direct costs inflicted by environmental degradation and the 'depreciation' of natural resources capital to allow for losses in future production potential. Although the national accounts record the income earned from harvesting resource stocks (such as fish catch, timber and minerals), the loss of future income through declining resource stocks and deteriorating environmental quality is excluded. By allowing for such 'depreciations' in the natural capital stock, the net contributions of resource degradation to national income are much lower, and more accurately reflect the impact on economic welfare (17). For example, Japan attempted to correct its national income figures for a variety of factors, including environmental ones. Accordingly, it has been found that instead of the GNP growing by a factor of 8.3 per cent per year between 1955 and 1985, it grew by an average of 5.8 per cent per year (14). In Indonesia, if the physical depletion, as well as net additions to petroleum, forest and soil assets are taken into consideration, it has been estimated that the GDP grew by 4.0 per cent per year in the period 1971 to 1984, instead of the reported gross value average of 7.1 per cent per year (18).

However, there are many difficulties in adjusting national accounts that remain to be solved. For one, measuring the stock of economic capital and its rate of depreciation is a complicated task in many developing countries. Some natural resources, such as

Classical economic theories and practices have treated nature as an infinite supply of physical resources to be used for human benefit, and as an infinite sink for the by-products of the development and consumption of these benefits, in the form of various types of pollution and ecological degradation.

soils and watersheds, are not easily measurable 'stocks' as such. Another problem is that the depreciation of natural resource stocks may not always include all the off-site environmental quality effects. For example, the total environmental costs of deforestation and timber extraction should include the economic costs of soil erosion, siltation of waterways, flooding and impacts on climate.

The changing world scene

The world has not been standing still while the debate on environment and development gathered momentum; ideas, concepts and issues emerged, were clarified and reiterated. The two decades since 1972 have seen major political, economic and social changes. The global political and economic landscape has altered, not gradually but in a number of dramatic and unforeseeable upheavals. As a result, the ideological and economic world maps of 1972 are no longer accurate in 1992; the geo-political assumptions which accompanied them do not hold true today; and the predictions of social change which were based on them have been proved inaccurate.

The challenge

The most dramatic and obvious political changes have been the most recent. The movement to democratic pluralism and the huge economic, social and political changes in the former Soviet Union, Central and Eastern Europe have gripped world attention and often dominated the news since the introduction of 'perestroika' in the mid-1980s. However, the causes are to be found earlier, in more subtle changes to the prevailing philosophies in both East and West which had more profound consequences than was immediately obvious. The change from an essentially bipolar world in which two super-powers and their supporters faced each other across an ideological and political abyss has created both opportunities and uncertainties. It may be some time before the geo-political 'post-perestroika' map is finally drawn, but the nature of that map and the world it represents will owe more to the fundamental causes of those changes than to the changes themselves.

The radical optimism of the early 1970s gave way under pressure of the global economic recession which followed the second oil shock in 1978. The belief, which under-pinned the Second UN Development Decade and the call for a 'New International Economic Order', that institutional solutions could be found to human and social problems, was replaced by a more individualistic, inward-looking,

Environmental issues may come to exercise a growing influence on international economic relations. They ... could influence the pattern of world trade, the international distribution of industry, the competitive position of different groups of countries, their comparative costs of production, etc.
Some environmental actions by developed countries ... are likely to have negative effects on developing countries' export possibilities and their terms of trade.
The Founex Report 1971 (2)

GATT, among other international organisations, could be used for the examination of the problems [of trade and the environment], specifically through the recently established Group on Environmental Measures and International Trade.
Stockholm Plan of Action 1972 (3)

Environmentally related regulations and standards should not be used for protectionist purposes.
UNEP 1987 (12)

market-oriented philosophy. Paradoxically, the same improvements in mass communications which have liberated individuals and fuelled popular demands for political reform have also led to an increased sense of individual helplessness in the face of mounting environmental crises, and greater popular distrust of politically-generated solutions to social, economic and environmental problems.

The few value-added products that are generated in developing countries are often blocked by lack of market access, as developing country commodity exports are affected by the 'new protectionism' which followed the recession of the early 1980s. Non-tariff barriers, voluntary export restraints, direct and indirect subsidies and other obstacles have made developing country access to northern markets extremely difficult. According to the World Bank, the percentage of OECD country imports covered by non-tariff barriers almost doubled between 1966 and 1986. Moreover, the percentage of trade affected by highly restrictive non-tariff measures is greater for developing countries than for industrialized countries. Subsidies on agricultural produce within the OECD are in the vicinity of US$300 billion per annum. The cost (in $1990) to the global South in 1980 of trade protectionism in developed countries has been estimated at around US$55 billion (19).

Over the past 20 years, both the World Bank and IMF have shifted development priorities from import substitution to export-led growth accompanied by severe structural adjustment programmes. For most developing countries with scant industrial capacities, there is little to export but natural resources, making these countries almost totally reliant on commodity exports. However, commodity prices have fallen steadily since the early 1970s. By 1986 average real commodity prices were at their lowest recorded levels in this century (with the single exception of 1932, the trough of the Great Depression). Prices of two critical export crops—cocoa and coffee—fell even further between 1986 and 1989 (by 48 and 55 per cent respectively). The World Bank forecasts that commodity prices are unlikely to rise during this decade, with intensified South-South competition in saturated markets.

The combined effect of debt servicing and reduced aid is a net financial flow from the South to the North. In 1989 developing countries paid US$59.5 billion in

The ideological and economic world maps of 1972 are no longer accurate in 1992; the geo-political assumptions which accompanied them do not hold true today; and the predictions of social change which were based on them have been proved inaccurate.

interest on their debts (World Bank, quoted in [20]), and received official development assistance of US$34.1 billion (17). In the same year the official debt of low- and middle-income countries grew by an average of 4 per cent (19). Increasing interest payments on a spiralling debt burden can be met only by increasing exports. For countries that are almost totally dependent on commodity exports in a hostile market, this means placing greater pressure on the environment and a further reduction in living standards for their people. With a projected one billion additional people sharing scarce resources in the global South in the near future, the pace of environmental degradation seems certain to increase unless the debt crisis is resolved and greater equity introduced to the world's commodity markets.

The opportunity

To concentrate only on the negative statistics of the past two decades and ignore promising present trends and recent events would give a distorted and overly negative picture. While we must accept the reality that the two decades since the Stockholm conference have seen a considerable degradation of the global environment and a further squandering of the world's stock of productive natural resources, there are also some grounds for optimism. A growing appreciation of the global nature of environmental problems and their implications—not just for the quality of life but for its very sustenance—has led to a new and more serious approach to environmental issues since the mid-1980s. Governments have displayed a greater willingness to act together to address environmental threats on a global basis, as was demonstrated by the successful negotiation between 1985 and 1987 of the Montreal Protocol on Substances that Deplete the Ozone Layer, its dramatic strengthening in 1990, and the large number of countries that have ratified it.

This rapid and decisive action (at least in terms of international treaty negotiations) and the steps taken since towards negotiation of conventions on control of hazardous wastes and their disposal, on climate change, and on biodiversity would have been hard to predict even a decade earlier. A willingness to act has been accompanied by an encouraging movement away from confrontation and towards a more cooperative approach by governments in forums dealing with environmental issues. It has thus been possible to develop new and innovative means (for example the funding mechanism established under the Montreal Protocol and the Global Environmental Facility) to address issues such as the transfer of environmentally sound technology to developing countries and to deal with major environmental problems.

International cooperation

Putting the world on the path of sustainable development will not be easy, given the environmental degradation and economic confusion that now prevail. The planning and implementation of development initiatives will have to change significantly, the global economy will have to be fundamentally restructured, and there will have to be a quantum leap in international cooperation. Unless the desire to ensure a sustainable future becomes a central concern of national governments, the continuing deterioration of the economy's natural support systems will eventually overwhelm efforts to improve the human condition.

The expectations for multinational cooperation raised at different forums over the past two decades have not been fulfilled. The global negotiations, whose immediate launching has been called for, have not materialized. The results of the Sixth United Nations Conference on Trade and Development have been disappointing for many, particularly the developing countries, and a similar disappointment has been felt over the failure to translate into concrete action the prescriptions for global economic recovery made at various summit meetings.

In the field of environment, the preparedness of governments to translate good intentions into action has been more positive. The trade in endangered species, wetlands and world heritage conventions, and the Montreal Protocol to protect the ozone layer (Chapter 21), have provided examples of major instruments for cooperation between and among developed and developing countries. But there is still an urgent need for the world community to solve a number of problems and to translate good intentions into practical actions to set the stage for sustainable and environmentally sound development.

There is, for example, mounting concern that conflicts are growing between international trade and environmental objectives. Many countries rely on imports of natural resources from developing countries that do not have alternative products to sell in

A willingness to act has been accompanied by an encouraging movement away from confrontation and towards a more cooperative approach by governments in forums dealing with environmental issues.

States have ... the sovereign right to exploit their own resources pursuant to their own environmental policies, and the responsibility to ensure that activities within their jurisdiction or control do not cause damage to the environment of other States or of areas beyond the limits of their national jurisdiction.
The Stockholm Declaration 1972 (3)

It cannot be too much or too often emphasised how crucial the dialogue between and amongst the developed and developing countries is for the appropriate solutions of environmental problems.
UNEP 1984 (22)

The responsibility for ensuring a better environment should be equitably shared and the ability of developing countries to respond be taken into account.
Langkawi Declaration 1989 (23)

If the multiple bonds that characterise interdependence are convincingly present in any field, it is in that encompassing development and the environment ... But the transition is not occurring smoothly and harmoniously; it is turbulent and beset with conflict.
Report of the South Commission 1990 (24)

international markets. For example, Malaysia, Indonesia, the Philippines, Côte d'Ivoire and Gabon supply some 80 per cent of the world market in tropical hardwoods. Yet it is clear that the tropical forests supplying these products are being used unsustainably (14). In so far as economic progress in the wealthier countries is sustainable, it could be said that the sustainability is in part being achieved by 'importing' it through unsustainability in other nations. Thailand exports its entire cassava production; 90 per cent of it goes to the European Community. This production, taking up about 1.5 million hectares of land, has led to the rapid degradation of natural resources in Thailand (15). Since 1982, the EC has tried to restrict its imports from Thailand, but this has not yet led to a decrease in production. In general, trade by industrialized countries and their trade-related policies have indirectly affected the environment and the use of natural resources in developing countries.

A number of factors have contributed to an unsustainable use of natural resources, soil degradation, excessive use of fertilizers and pesticides, and pollution in many developing countries. They include developing countries' debt and/or balance of payment problems, industrialized countries' protectionism against goods manufactured in developing countries, preferential treatment of raw materials from developing countries, domestic agricultural subsidies in a number of developed countries and price fluctuations on the world market (15). There are also fears that the recent trends in international trade liberalization may have considerable negative consequences for the environment.

Twenty years ago development assistance agencies and financial institutions gave little attention to environmental protection. Recently, however, many of these agencies and institutions have established formal procedures for assessing the environmental impacts of their development assistance activities. Although these measures are welcomed to help developing countries chart their future development in an environmentally sound manner, it is feared that they will constitute a new 'conditionality' on providing development and/or technical assistance. It should be emphasized here that every country has the sovereign right to manage its own natural resource base, as well as the responsibility to protect its own environment and to ensure that its development activities do not harm the environment of its neighbours.

Priorities for the next two decades

Environmental problems cut across a range of policy issues and are mostly rooted in inappropriate development patterns. Consequently, environmental issues, goals and actions cannot be framed in isolation from the development and policy sectors from which they emanate. Against this background, the Stockholm Conference produced an Action Plan for the Human Environment which was endorsed in General Assembly Resolution 2994 (XXVII) of 15 December 1972. The 109 recommendations in the Plan fell into three groups concerned with environmental assessment (evaluation and review, research, monitoring and information exchange), environmental management and supporting measures (education and training, public information, financing and technical co-operation).

In the past two decades, the Stockholm recommendations constituted the basis for action by the UN system and other international bodies. Over the years specific goals were set up to implement these recommendations. These goals were revised and refined as our scientific knowledge of the different issues evolved and improved. This process led to the formulation, and adoption, by the General Assembly of the United Nations in 1987 (GA resolution 42/186 of 11 December 1987), of the *Environmental Perspective to the Year 2000 and Beyond*. This document reflects an intergovernmental consensus on growing environmental challenges to the year 2000 and beyond, in six major sectors: population, food and agriculture, energy, industry, health and human settlements, and international economic relations. In addition, the document discusses briefly four issues of global concern (oceans and seas, outer space, biological diversity, and security and environment) and considers the different instruments of environmental action.

Also welcomed by the General Assembly (GA resolution 42/187 of 11 December 1987) was *Our Common Future,* the report of the World Commission on Environment and Development. This document addressed specifically the need for sustainable development and the legal principles on which it should be based.

Taking into consideration the recommendations made and priorities outlined in various documents, including those

Unless the desire to ensure a sustainable future becomes a central concern of national governments, the continuing deterioration of the economy's natural support systems will eventually overwhelm efforts to improve the human condition.

mentioned in this chapter, and the goals and targets presented to the Governing Council of UNEP in 1987, the time is now ripe to sharpen the focus on a number of issues that should be addressed by the world community in the coming two decades. Some specific, achievable actions to translate words into deeds are outlined in the box on pages 258–259. The targeted actions proposed for consideration in this box do not constitute an exhaustive list. Nor are they intended merely to address the symptoms they relate to. They provide a practical basis for direct action for environmental improvement, and for the design and implementation of national and international policies and programmes to reconcile social, economic

Priorities for action

Regulatory measures

By 1995:

- A global agreement on reforestation targets for each decade of the 21st century in each of the world's eco-regions.
- A global plan to combat marine pollution from land-based sources, with a target to reduce by the year 2000 marine pollution from such sources to the 1990 level, and an agreed programme for further reductions after 2000.
- International agreement to ban all exports of hazardous wastes to developing countries, and a timetable to reduce the generation of such wastes.
- A global convention for the exchange of information on chemicals in international trade and establishment of an intergovernmental mechanism for chemical risk assessment and management.
- A global convention on the prevention, notification and mutual cooperation in mitigating the effects of major environmental emergencies.
- An international code of conduct to apply internationally agreed guidelines for the transfer of technology, particularly to developing countries.
- An international agreement on the guidelines for application of environmental impact assessment especially with regard to human activities with potential transboundary effects.
- Establish an international non-governmental body to help in monitoring breaches of environmental treaties and national actions leading or likely to lead to major environmental deterioration.

By 2000:

- Agreement on the means of ensuring compliance with environmental treaties and establishment of appropriate institutional mechanisms to verify their implementation.

Assessment

By 1995:

- Assess the environmental impacts of known new and alternative sources of energy.

By 2000 complete the following:

- Comprehensive assessment of air quality in all urban areas.
- Comprehensive assessment of global freshwater resources and their quality.
- Comprehensive assessment of land and soil degradation in the world.
- Environmental impact assessment of existing new technologies.
- Environmental impact assessment of existing new materials.

By 2010:

- Complete a survey of the world's habitats known to be unique, rich in biodiversity, or at risk.

and environmental objectives in development. They represent goals that can be achieved through integrated development planning which addresses the underlying causes of environmental degradation and lack of human development: unmanageable population growth, grinding poverty, crushing debt and unfair international economic relations on the one hand; and unsustainable lifestyles, unnecessary over-consumption and irresponsible use of scarce human and financial resources on the other. These priorities are based on pronouncements already made by governments, documents noted by them, and existing studies, publications and global strategies such as the *World Conservation Strategy* (25) and *Caring For The Earth* (26).

Environmental management

By 1995:

- Approve three concrete five-year programmes to combat land degradation in drylands (desertification), which have been costed and the sources of funding identified.
- Establish an international programme to improve the efficiency of irrigation systems with the goal of reducing wastage of irrigation water by 10 per cent each decade.
- Establish a UN Centre for Response to Environmental Emergencies.

By 2000:

- Achieve a 30 per cent reduction in the amount of hazardous waste generated, compared with the 1990 level.
- Adoption by major development financing institutions of policies and procedures that ensure that their financial support to activities does not lead to environmental deterioration.
- All countries to adopt environmental and natural resource accounting as part of their system of national accounts.
- Capital flows in the form of natural resource imports and exports to be included in international trade statistics.
- All countries with real GDP per capita above US$5000 to produce a plan to reduce their consumption of non-renewable natural resources.
- All countries with per capita annual energy consumption over 80 gigajoules to stabilize consumption at 1992 rates and establish programmes to reduce energy use to the 80 gigajoule level.

By 2000–2010:

- End net global deforestation.

Environment and economics

By 1995:

- Estimates of the global costs of failing to deal with climate change, ozone layer depletion, loss of biodiversity, marine and coastal deterioration from land-based sources of pollution, and continued production of hazardous wastes.
- Production of revised estimates of the additional resources needed for the transfer of knowledge, information and specific environmentally sound technologies to developing countries and countries in transition, to allow them to participate meaningfully in dealing with their national as well as global environmental problems. By the same year, agreement should be reached on sources of funding and mechanisms for the transfer of technologies.

References

Chapter 1

(1) OECD (1991) *The State of the Environment—1991,* OECD, Paris.

(2) Shah, J.J. and Singh, H.B. (1988) Distribution of volatile organic chemicals in outdoor and indoor air. *Environmental Science and Technology,* vol 22, p 1381.

(3) GEMS/WHO (1988) *Assessment of Urban Air Quality Worldwide,* WHO, Geneva.

(4) UNEP (1989) *Environmental Data Report,* Blackwell, Oxford.

(5) French, H.F. (1990) *Clearing the Air: a global agenda,* Worldwatch Paper 94, Worldwatch Institute, Washington, D.C.

(6) Sakugawa, H. *et al* (1990) Atmospheric hydrogen peroxide. *Environmental Science and Technology,* vol 24, p 1452.

(7) WHO (1990) *Indoor air quality: biological contaminants,* WHO Regional Publication, European Series No 31, WHO, Geneva.

(8) Moseley, C. (1990) Indoor air quality problems. *Journal of Environmental Health,* vol 53, p 19.

(9) NRC (1981) *Indoor Pollutants,* National Research Council, National Academy Press, Washington, D.C.

(10) Spengler, J. *et al* (1982) Indoor air pollution. *Environment International,* Special Issue, 8.

(11) Spengler, J. and K. Sexton (1983) Indoor air pollution: a public health perspective. *Science,* vol 221, p 9.

(12) Nero, A.V. (1988) Controlling indoor air pollution. *Scientific American,* vol 258, p 24.

(13) WHO (1989) *Indoor air quality: organic pollutants,* WHO, European Reports and Studies 111, WHO, Geneva.

(14) Davidson, C.I. *et al* (1986) Indoor and outdoor air pollution in the Himalayas. *Environmental Science and Technology,* vol 20, p 561.

(15) Rodhe, H. *et al* (1988) *Acidification in Tropical Countries,* SCOPE 36, John Wiley, Chichester.

(16) UNEP (1991) *Environmental Data Report,* Blackwell, Oxford.

(17) Gaffney J.S. *et al* (1987) Beyond acid rain. *Environmental Science and Technology,* vol 21, p 519.

(18) Madadevan T.N. *et al* (1986) Trace elements in precipitation: over an industrial area of Bombay. *The Science of the Total Environment,* vol 48, p 213.

(19) Lum, K.R. *et al* (1987) Bioavailable Cd, Pb and Zn in wet and dry deposition. *The Science of the Total Environment,* vol 63, p 161.

(20) Hileman, B. (1983) Acid fog. *Environmental Science and Technology,* vol 17, p 117A.

(21) Jacob D.J. *et al* (1985) Chemical composition of fogwater collected along the California coast. *Environmental Science and Technology,* vol 19, p 730.

(22) Young, J.R. *et al* (1988) Deposition of airborne acidifiers in the Western environment. *Journal of Environmental Quality,* vol 17, p 1.

(23) Glotfelty D.E. *et al.* (1987) Pesticides in fog. *Nature,* vol 325, p 602.

(24) Dassen, W. *et al* (1986) Decline in Childrens' pulmonary function during an air pollution episode. *Journal of Air Pollution Control Association,* vol 35, p 1223.

(25) Graedel, T.E. and McGill, R. (1986) Degradation of materials in the atmosphere. *Environmental Science and Technology,* vol 20, p 1093.

(26) Nazaroff W.W. and Teichmann, K. (1990) Indoor radon. *Environmental Science and Technology,* vol 24, p 774.

(27) WHO (1984) *Biomass Fuel Combustion and Health,* Report EEP/84, 64. WHO, Geneva.

(28) Smith, K.R. (1986) Biomass combustion and indoor air pollution. *Environmental Management,* vol 10, p 61.

(29) UNEP (1987) *The State of the Environment,* UNEP, Nairobi.

(30) McIlvaine, R.W. (1991) The 1991 global air pollution control industry. *Journal Air and Waste Management Association,* vol 41, p 272.

(31) McCormick, J. (1985) *Acid Earth,* Earthscan, London.

(32) OECD (1985) *The State of the Environment—1985,* OECD, Paris.

(33) Dignon, J. and Hanseed, S. (1989) Global emissions of nitrogen and sulphur oxides from 1860 to 1980. *Journal Air Pollution Control Association,* vol 39, p 180.

(34) Nriagu, J.O. (1988) Quantitative assessment of worldwide contamination of air, water and soils by trace elements. *Nature,* vol 333, p 134.

(35) Nriagu, J.O. (1989) A global assessment of natural sources of atmospheric trace metals. *Nature,* vol 338, p 47.

Chapter 2

(1) Crutzen, P.J. (1971) Ozone production rates in an oxygen, hydrogen and nitrogen-oxides atmosphere. *Journal Geophysical Research,* vol 76, p 7311.

(2) Johnston, H.S. (1971) Reduction of stratospheric ozone by nitrogen oxide catalysts from supersonic transport exhaust. *Science,* vol 173, p 517.

(3) Molina, M.J. and Rowland, F.S. (1974) Stratospheric sink for chlorofluoromethane. *Nature,* vol 249, p 810.

(4) Rowland, F.S. and Molina, M.J. (1975) Chlorofluoromethanes in the environment. *Revue of Geophysics and Space Physics,* vol 13, p 1.

(5) Rowland, F.S. (1987) Can we close the ozone hole? *Technology Review,* August/September, p 51.

(6) Rowland, F.S. (1990) Stratospheric ozone depletion by chlorofluorocarbons. *Ambio,* vol 19, p 281.

(7) Rowland, F.S. (1991) Stratospheric ozone in the 21st century. *Environmental Science and Technology,* vol 25, p 622.

(8) Prather, M.J. *et al* (1984) Reductions in ozone at high concentrations of stratospheric halogens. *Nature,* vol 312, p 227.

(9) NAS (1979) *Stratospheric ozone depletion by halocarbons,* National Academy

Press, Washington, D.C.

(10) NASA (1979) *The Stratosphere: present and future,* NASA Reference Publication 1049, NASA Goodard Space Flight Centre.

(11) DOE (1979) *Chlorofluorocarbons and their effects on stratospheric ozone,* Pollution Paper No 15, UK Department of Environment, Her Majesty's Stationery Office, London, UK.

(12) NRC (1984) *Causes and Effects of Changes in Stratospheric Ozone: Update 1983,* National Research Council, National Academy Press, Washington, D.C.

(13) WMO/NASA (1985) *Atmospheric Ozone—1985,* Global Ozone Research & Monitoring Project, Report No 16, 3 volumes, WMO, Geneva.

(14) UNEP (1986) *Report of the 8th Session of the Coordinating Committee on the Ozone Layer,* UNEP/CCOL/VIII, UNEP, Nairobi.

(15) Watson, R.J. (1988) *Current scientific understanding of stratospheric ozone,* UNEP/ozl. sc. 1/3, UNEP, Nairobi.

(16) Farman, J.C. *et al* (1985) Large losses of total ozone in Antarctica reveal ClO_x-NO_x interaction. *Nature,* vol 315, p 207.

(17) Bowman, K.P. (1988) Global trends in total ozone. Science, vol 239, p 48.

(18) WMO (1989) *Scientific Assessment of Stratospheric Ozone 1989,* Global Ozone Research and Monitoring Project, Report No 20 , WMO, Geneva.

(19) UNEP (1989) *Reports of the Ozone Scientific Assessment, Economic and Environmental Effects Panels,* UNEP, Nairobi.

(20) NASA (1991) Statement by R.T. Watson and R.S. Stolarski before the subcommittee on Science, Technology and Space of the Committee on Commerce, Science & Transportation of the U.S. Senate's 102nd Conference.

(21) Pitcher, H.M. and Longstreth, J.D. (1991) Melanoma mortality and exposure to ultraviolet radiation. *Environment International,* vol 17, p 7.

(22) Dudek, D.J. *et al* (1990) Cutting the cost of environmental policy: lessons from business response to CFC regulation. *Ambio,* vol 19, p 324.

(23) Bruhl, C. and Crutzen, P J. (1990) Ozone and climate change in the light of the Montreal Protocol. *Ambio,* vol 19, p 293.

(24) Rosemarin, A. (1990) Some background on CFCs. *Ambio,* vol 19, p 220.

Chapter 3

(1) SCEP (1970) *Man's impact on the global climate: Report of the Study of Critical Environmental Problems,* The MIT Press, Cambridge, Mass. 319 pp.

(2) SMIC (1971) *Inadvertent Climate Modification: Report of the Study of Man's Impact on Climate,* The MIT Press, Cambridge, Mass. 308 pp.

(3) WMO (1979) *Proceedings of the World Climate Conference,* Report No 537, WMO, Geneva.

(4) Neftel, A. (1985) Evidence from polar ice cores for the increase in atmospheric carbon dioxide in the past two centuries. *Nature,* vol 315, p 45.

(5) IPCC (1990) *Climate Change,* Report of the Intergovernmental Panel on

Climate Change, WMO/UNEP, Cambridge University Press. See also IPCC Working Group II & III, WMO/UNEP, UNEP, Nairobi.

(6) Houghton R.A. (1990) The global effects of tropical deforestation. *Environmental Science and Technology,* vol 24, p 414.

(7) Ehhalt, D.H. (1985) Methane in the global atmosphere. *Environment,* vol 27, p 6.

(8) Rasmussen R.A. and Khalil, M.A. (1986) The behaviour of trace gases in the troposphere. *The Science of the Total Environment,* vol 48, p 169.

(9) Mooney H.A. *et al* (1987) Exchange of materials between terrestrial ecosystems and the atmosphere. *Science,* vol 238, p 926.

(10) Khalil M.A. *et al* (1991) Methane emissions from rice fields in China. *Environmental Science and Technology,* vol 25, p 979.

(11) Manabe S. and Wetherald, R.T. (1967) Thermal equilibrium of the atmosphere with a given distribution of relative humidity. *Journal of Atmospheric Science,* vol 24, p 241.

(12) Schlesinger M.E. and Mitchell, J.F.B. (1985) Model projections of equilibrium response to increased CO_2 concentration, in *Projecting the climatic effects of increasing carbon dioxide,* Report DOE/ER-0237, US Department of Energy, Washington, D.C., 81–148.

(13) UNEP/ICSU/WMO (1986) *Report of the International Conference on the Assessment of the Role of Carbon Dioxide and of other Greenhouse Gases in Climate Variations and Associated Impacts,* WMO Report 661, WMO, Geneva.

(14) MacDonald, G. (1989) Scientific basis for the greenhouse effect, in *The Challenge of Global Warming* (ed D.E. Abrahamson), Island Press, Washington, D.C.

(15) Johnes P. *et al* (1986) Global temperature variation between 1861 and 1984. *Nature,* vol 322, p 430.

(16) Wirth D.A. and Lashof, D.A. (1990) Beyond Vienna and Montreal—Multilateral agreements in greenhouse gases. *Ambio,* vol 19, p 305.

Chapter 4

(1) UNEP (1990) *Technical annex to the report on the state of the marine environment,* UNEP Regional Seas Reports and Studies 114/1, UNEP, Nairobi.

(2) GESAMP (1990) *The State of the Marine Environment,* Regional Seas Reports and Studies No 115, UNEP, Nairobi.

(3) Arnando, R. (1990) *The problem of persistent plastics and marine debris in the oceans,* UNEP Regional Seas Reports and Studies No 114/1, UNEP, Nairobi.

(4) NRC (1985) *Oil in the Sea: Inputs, Fates and Effects,* National Academy Press, Washington, D.C.

(5) IMO (1990) *International Maritime Organization,* Briefing IMO/810/90, IMO, London.

(6) Gosselin, S. *et al* (1989) Vulnerability of marine fish larvae to the toxic dinoflagellate. *Marine Ecology Progress Series,* vol 57, p 1.

(7) WRI (1990) *World Resources 1990-1991,* Oxford University Press, New York.

(8) OECD (1991) *The State of the Environment—1991,* OECD, Paris.

(9) Lean, G. *et al* (1990) *Atlas of the Environment,* Arrow Books, London.

(10) Hinrichsen, D. (1990) *Our Common Seas: Coasts in Crisis,* Earthscan Publications, London.

(11) FAO (1988) *Country Tables,* FAO, Rome.

(12) FAO (1991) *The State of Food and Agriculture—1990,* FAO, Rome.

(13) Angel, M.V. (1987) Criteria for protected areas and other conservation measures in the Antarctic region. *Environment International,* vol 13, p 105.

(14) Crockett, R.N. and Clarkson, P.D. (1987) The exploitation of Antarctic minerals. *Environment International,* vol 13, p 121.

(15) Mitchell, B. (1988) Undermining Antarctica. *Technology Review,* Feb/March, p 51.

(16) UNEP (1991) *The State of the Environment,* UNEP, Nairobi.

Chapter 5

(1) White, G.F. (1988) A century of change in world water management, in *Proceedings of the Centennial Symposium—Earth '88,* National Geographic Society, Washington, D.C., p 248.

(2) WRI (1987) *World Resources—1987*, Basic Books, New York.

(3) La Riviere, J.W. (1989) Threats to the world's water. *Scientific American,* vol 261, p 48.

(4) Shiklomanov, I.A. (1986) Water consumption, water availability and large-scale water projects in the world, in *International Symposium on the Impact of Large Water Projects on the Environment,* UNESCO, Paris.

(5) GEMS/WHO (1989) *Global Freshwater Quality: a first assessment.* Blackwell, Oxford.

(6) OECD (1991) *The State of the Environment—1991.* OECD, Paris.

(7) El-Hinnawi, E. (1991) *Sustainable agriculture and rural development in the Near East,* Regional Document No 4, FAO/Netherland Conference on Agriculture & Environment, FAO, Rome.

(8) Falkenmark, M. (1986) Fresh waters as a factor in strategic policy and action, in *Global Resources and International Conflict* (ed A.H. Westing), Oxford University Press, Oxford.

(9) Biswas, A. *et al* (1983) *Long-Distance Water Transfer,* Tycooly International, Dublin.

(10) ICOLD (1989) *World Register of Dams—1988 updating.* International Commission on Large Dams, Paris.

(11) Veltrop, J.A. (1991) Water, dams and hydropower in the coming decades. *Water Power and Dam Construction,* June 1991, p 37.

(12) El-Hinnawi, E. (1981) *Environmental Impacts of Production and Use of Energy,* Tycooly International, Dublin.

(13) El-Hinnawi, E. and Biswas, A. (1981) *Renewable Sources of Energy and Environment,* Tycooly International, Dublin.

(14) White, G.F. (1988) The environmental effects of the High Dam at Aswan. *Environment,* vol 30, p 5.

(15) United Nations (1990) *Achievements of the International Drinking Water Supply and Sanitation Decade, 1981-1990,* Report of the Secretary General A/45/327, United Nations, New York.

(16) Najlis, P. and Edwards, A. (1991) The international drinking water supply and sanitation decade in retrospect and implications for the future. *Natural Resources Forum,* May 1991, p 110.

(17) CNRET (1978) Register of International Rivers. *Water Supply and Management,* vol 2, p 1.

Chapter 6

(1) FAO (1990) *FAO Production Yearbook,* vol 43, FAO, Rome.

(2) Weaver C.E. (1989) *Clays, muds and shales,* Elsevier Science Publishers, Amsterdam.

(3) UNEP/FAO (1983) *Guidelines for the control of soil degradation,* FAO, Rome.

(4) WRI (1986) *World Resources—1986,* Basic Books, New York.

(5) Brown L.R. and Wolf, E.C. (1984) *Soil Erosion: Quiet Crisis.* Worldwatch Paper No 60, Worldwatch Institute, Washington, D.C.

(6) Eckholm, E. (1976) *Losing Ground,* W.W. Norton, New York.

(7) Jalees, K. (1985) Loss of productive soil in India. *International Journal of Environmental Studies,* vol 24, p 245.

(8) ISRIC (1990) *World status of human-induced soil degradation,* International Soil Reference and Information Centre, Wageningen, Netherlands.

(9) UNEP (1991) *Status of Desertification and Implementation of the UN Plan of Action to Combat Desertification,* UNEP, Nairobi.

(10) El-Hinnawi, E. (1985) *Environmental Refugees,* UNEP, Nairobi.

(11) Pavlov, G. (1982) *The Protection and Improvement of the Environment in Bulgaria,* CMEA, Committee for Scientific and Technological Cooperation, Moscow.

(12) ESCAP (1990) *The State of the Environment in the ESCAP region,* ESCAP, Bangkok.

(13) El-Hinnawi, E. (1991) *Sustainable agriculture and rural development in the Near East,* Regional document No 4, FAO/Netherlands Conference on Agriculture and Environment, FAO, Rome.

Chapter 7

(1) FAO (1991) *Protection of Land Resources: deforestation,* Prepcom UNCED, 2nd Session. Document A/CONF.151/PC/27.

(2) OECD (1991) *The State of the Environment—1991,* OECD, Paris.

(3) UNEP (1991) *Environmental Data Report,* 3rd edn, Blackwell, Oxford.

(4) El-Hinnawi, E. and Hashmi, M. (1987) *The State of the Environment,* Butterworths, London.

(5) ECE/UNEP (1989) *Forest Damage and Air Pollution,* Report of the 1988 forest damage survey in Europe, ECE, Geneva.

(6) ECE (1990) *The State of the Transboundary Air Pollution: 1989 Update,* ECE/EB.AIR/25, ECE, Geneva.

(7) ECE (1984) *Airborne Sulphur Pollution,* Air Pollution Studies No 1, ECE, Geneva.

(8) Blank, I.W. (1985) A new type of forest decline in Germany. *Nature,* vol 314, p 311.

(9) Binns, W.O. (1985) Effects of acidic deposition on forests and soils. *The Environmentalist,* vol 5, p 279.

(10) Nihlgard, B. (1985) The ammonium hypothesis is an additional explanation for the forest dieback in Europe. *Ambio,* vol 14, p 2.

(11) FAO/UNEP (1982) *Tropical Forest Resource,* FAO, Rome.

(12) El-Hinnawi, E. (1985) *Environmental Refugees,* UNEP, Nairobi.

(13) Lean, G. *et al* (1990) *Atlas of the Environment,* Arrow Books, London.

(14) Corson, W.H. (1990) *The Global Ecology Handbook,* Beacon Press, Boston.

(15) Houghton, R.A. (1990) The global effects of tropical deforestation. *Environmental Science and Technology,* vol 24, p 414.

(16) Myers, N. (1989) The future of forests, in *The Fragile Environment* (eds L. Friday and R. Laskey), Cambridge University Press.

(17) IPCC (1990) *Potential Impacts of Climate Change,* Report of Working Group II, Intergovernmental Panel on Climate Change, WMO, Geneva.

(18) El-Hinnawi, E. (1991) *Sustainable agriculture and rural development in the Near East,* Regional document No 4, FAO/Netherlands Conference on Agriculture and Environment, FAO, Rome.

(19) Harlow, C.S. and Adriano, A.S. (1980) The Philippine dendrothermal power programme, in *Proceedings of Bioenergy '80 Congress,* Bioenergy Council, Washington, D.C.

(20) El-Hinnawi, E. and Biswa, A. (1981) *Renewable Sources of Energy and the Environment.* Tycooly International, Dublin.

(21) Gradwohl, J. and Greenberg, R. (1988) *Saving the Tropical Forests,* Earthscan Publications Ltd., London.

(22) George, S. (1988) *A Fate Worse Than Debt,* Penguin Books, London.

Chapter 8

(1) Wilson, E.O. (1988) *Biodiversity,* National Academy Press, Washington, D.C.

(2) McNeely, J.A. *et al* (1990) *Conserving the World's Biological Diversity*, IUCN, Gland.

(3) Reid, W.V. and Miller, K.R. (1989) *Keeping Options Alive: the scientific basis for conserving biodiversity*, World Resources Institute, Washington, D.C.

(4) Gentry, A.H. (1988) Tree species richness of upper Amazonian forests. *Proceedings US National Academy of Science*, vol 85, p 156.

(5) Davis, S.D. *et al* (1986) *Plants in Danger: what do we know*, IUCN, Gland.

(6) Dugan, P.J. (1990) *Wetland Conservation: a review of current issues and required action*, IUCN, Gland.

(7) WRI (1990) *World Resources 1990–1991*, Oxford University Press, New York.

(8) Ehrlich P. and Ehrlich A. (1982) *Extinction: the causes and consequence of the disappearance of species*, Victor Gollancz, London.

(9) Raven P.H. (1988) Our diminishing tropical forests, in *Biodiversity*, (ed E.O. Wilson), National Academy Press, Washington, D.C.

(10) WCMC (1992) *Global Biodiversity—1992: status of the earth's living resources*, World Conservation Monitoring Centre, Cambridge, in press.

(11) Lean, G. *et al* (1990) *Atlas of the Environment*, Arrow Books, London.

(12) Prescott-Allen, C. and Prescott-Allen, R. (1986) *The First Resource: wild species in the North American economy*, Yale University Press, New Haven, CT.

(13) IUCN/UNEP/WWF (1991) *Caring for the Earth*, IUCN, Gland.

(14) UNEP (1991) *Environmental Data Report*, 3rd edn, Blackwell, Oxford.

Chapter 9

(1) OFDA (1987) *Disaster History: major disasters worldwide, 1900–present*, Office of Disaster Assistance, USAID, Washington, D.C.

(2) Berz, G.A. (1991) Global warming and the insurance industry. *Nature and Resources*, UNESCO, vol 27, p 19.

(3) Berresheim, H. and Jaeschke, W. (1983) The contribution of volcanoes to the global atmospheric sulphur budget. *Journal of Geophysical Research*, vol 88, C6, p 3732.

(4) Bimblecombe, P. and Lein, A.Y. (1989) *Evolution of the Global Biogeochemical Sulphur Cycle*, SCOPE 39, John Wiley & Sons, Chichester.

(5) Pollack, J.B. *et al* (1976) Volcanic explosions and climate change. *Journal of Geophysical Research*, vol 81, p 1071.

(6) Robock, A. (1978) Internally and externally caused climate change. *Journal of Atmospheric Science*, vol 35, p 1111.

(7) Toon, O.B. and Pollack, J.B. (1982) Stratospheric aerosols and climate, in *The Stratospheric Aerosol Layer* (ed R.C. Witten), Springel, Berlin.

(8) NRC (1985) *The Effects on the Atmosphere of a Major Nuclear Exchange*, National Academy Press, Washington, D.C.

(9) El-Hinnawi, E. (1981) *The Environmental Impacts of Production and Use of Energy*, Tycooly International, Dublin.

(10) El-Hinnawi, E. and Biswas, A. (1981) *Renewable Sources of Energy and Environment*, Tycooly International, Dublin.

(11) *New Scientist*, 24 August 1991.

(12) OECD (1991) *The State of the Environment—1991*, OECD, Paris.

(13) UNEP (1991) *Environmental Data Report*, 3rd edn, Blackwell, Oxford.

(14) WCED (1987) *Our Common Future*, Oxford University Press, Oxford.

(15) Wilhite, D.A. (1990) The enigma of drought: management and policy issues for the 1990s. *International Journal of Environmental Studies*, vol 36, p 41.

(16) Glantz, M. *et al* (1987) *The societal impacts associated with the 1982-83 worldwide climate anomalies*, National Centre for Atmospheric Research, Boulder, Colorado, and UNEP, Nairobi.

(17) Trenberth, K.E. *et al* (1988) Origins of the 1988 North American drought. *Science*, vol 242, p 1640.

(18) WMO (1986) *Report on drought and countries affected by drought during 1974–1985*, WCP-118, WMO/TD No 133, WMO, Geneva.

(19) Mattson, S.J. and Haack, R.A. (1987) The role of drought in outbreaks of plant-eating insects. *Bioscience*, vol 37, p 110.

(20) El-Hinnawi (1985) *Environmental Refugees*, UNEP, Nairobi.

(21) Velez, R. (1990) Mediterranean forest fires: a regional perspective. *Unasylva*, vol 41, p 3.

(22) McCleese, W.L. *et al* (1991) Real-time detection, mapping and analysis of wildland fire information. *Environment International*, vol 17, p 111.

(23) Ward, D.E. and Hardy, C.C. (1991) Smoke emissions from wildland fires. *Environment International*, vol 17, p 117.

(24) *Oil Spill Intelligence Report* (1991) vol XIV, No 12.

(25) IMO (1990) Feature published 1990, IMO, London.

(26) NRC (1985) *Oil in the sea*, National Academy Press, Washington, D.C.

(27) GESAMP (1990) *The State of the Marine Environment*, UNEP, Nairobi.

(28) Smets, H. (1988) The cost of accidental pollution. *Industry and environment*, vol 11, p 28.

(29) Cohen, M.A. (1986) The costs and benefits of oil spill prevention and enforcement. *Journal of Environmental Economics and Management*, vol 13, p 167.

(30) Maki, A.W. (1991) The Exxon Oil Spill: initial environmental impact assessment. *Environmental Science and Technology*, vol 25, p 24.

(31) Kelso, D.D. and Kendziorek, M. (1991) Alaska's response to the Exxon Valdez oil spill. *Environmental Science and Technology*, vol 25, p 16.

(32) Hay, A.W.M. (1977) Tetrachlorodibenzo-p-dioxin release at Seveso. *Disasters*, vol 1, p 289.

(33) Hay, A. (1978) Seveso: no answers yet. *Disasters*, vol 2, p 163.

(34) Cardillo, P. *et al* (1984) The Seveso case and the safety problem in production of 2,4,5 trichlorophenol. *Journal of Hazardous Materials*, vol 9, p 221.

(35) Bowonder, B. (1985) The Bhopal accident: implications for developing countries. *The Environmentalist*, vol 5, p 89.

(36) Bowonder, B. (1987) An analysis of the Bhopal accident. *Project Appraisal*, vol 2, p 157.

(37) Weir, D. (1987) *The Bhopal Syndrome*, Earthscan Publications, London.

(38) Capel, P.D. *et al* (1988) Accidental input of pesticides into the Rhine river. *Environmental Science and Technology*, vol 22, p 992.

(39) ILO (1991) *Prevention of Major Industrial Accidents*, ILO, Geneva.

(40) IAEA (1991) Nuclear power status around the world. *IAEA Bulletin*, vol 33, p 43.

(41) Franzen, F. (1987) Reviewing the operational safety of nuclear power plants. *IAEA Bulletin*, vol 4, p 13.

(42) Scott, R.L. and Gallaher, R.B. (1976) Recent occurences at nuclear reactors and their causes. *Nuclear Safety*, vol 17, p 611.

(43) Mays, G.T. and Gallaher, R.B. (1982) Events resulting in reactor shutdown and their causes. *Nuclear Safety*, vol 23, p 85.

(44) Mays, G.T. and Gallaher, R.B. (1983) Events resulting in reactor shutdown and their causes. *Nuclear Safety*, vol 24, p 250.

(45) Silver, E.G. (1984) Reactor shutdown experience. *Nuclear Safety*, vol 25, p 834.

(46) Swaton, E. *et al* (1987) Human factors in the operation of nuclear power plants. *IAEA Bulletin*, vol 29, p 27.

(47) WASH-1400 (1975) *Reactors safety study*, US Atomic Energy Commission, Washington, D.C.

(48) Lewis, H.W. *et al* (1978) *Risk assessment review group*, Report NUREG/CR-400, Washington, D.C.

(49) NAS (1979) *Risks Associated with Nuclear Power*, National Academy of Science, Washington, D.C.

(50) El-Hinnawi, E. (1980) *Nuclear Energy and Environment*, Pergamon Press, Oxford.

(51) Scott, R.L. (1976) Browns Ferry Nuclear Power Plant Fire on March 22, 1975. *Nuclear Safety*, vol 17, p 592.

(52) Toth, L.M. *et al* (1986) The Three Mile Island Accident. *American Chemical Society Symposium Series 293*, Washington, D.C.

(53) Ilyin, L.A. and Pavlovskij, O.A. (1987) Radiological consequences of the Chernobyl accident. *IAEA Bulletin*, vol 29, p 17.

(54) Anspanch, L.R. *et al* (1988) The global impact of the Chernobyl reactor accident. *Science*, vol 242, p 1513.

(55) Hohenemser, C. and Renn, O. (1988) Chernobyl's other legacy. *Environment*, vol 30, p 5.

(56) Jaworowski, Z. (1988) Chernobyl proportions. *Environment International*, vol 14, p 69.

(57) Bennet, B. and Bouville, G. (1988) Radiation doses in countries of the northern hemisphere from the Chernobyl nuclear reactor accident. *Environment International*, vol 14, p 75.

(58) Savchenko, V.K. (1991) The Chernobyl catastrophe and the biosphere. *Nature and Resources*, UNESCO, vol 27, p 37.

(59) Webb, J. (1991) Chernobyl findings. *New Scientist*, 1 June, p 17.

(60) Rich, V. (1991) An ill wind from Chernobyl. *New Scientist*, 20 April, 26–28.

(61) Bojcun, M. (1991) The legacy of Chernobyl. *New Scientist*, 20 April, 30–36.

(62) Flavin, C. (1987) Reassessing Nuclear Power, in *State of the World,* (eds L.R. Brown *et al*), W.W. Norton & Co., New York.

(63) Wilkins, L. (1987) *Share vulnerability: media coverage and public memory of the Bhopal disaster*, Westport, Conn., Greenwood Press.

(64) Wilkins, L. and Patterson, P. (1987) Risk analysis and the construction of news. *Journal of Communication*, vol 37, p 80.

(65) Hazarika, S. (1987) *Bhopal, the lesson of a tragedy*, Penguin, London.

(66) Friedman, S.M. *et al* (1987) Reporting on Radiation: a content analysis of Chernobyl coverage. *Journal of Communication*, vol 37, p 58.

(67) IAEA (1988) Radiation Sources: lessons from Goiania. *IAEA Bulletin*, vol 30(4), p 10.

(68) Oliver-Smith, A. (1991) Successes and failures in post disaster resettlement. *Disasters*, vol 15, p 12.

(69) Pearce, F. (1991) Acts of God, acts of man? *New Scientist*, 18 May, 20–21.

Chapter 10

(1) NRC (1984) *Toxicity Testing*, National Academy Press, Washington, D.C.

(2) Crump, A. (1991) *Dictionary of Environment and Development*, Earthscan Publications, London.

(3) Glotfelty, D.E. *et al* (1987) Pesticides in Fog. *Nature*, vol 325, p 602.

(4) Travis, C.C. and Hester, S.T. (1991) Global chemical pollution. *Environmental Science and Technology*, vol 25, p 814.

(5) Bidelman, T.F. (1988) Atmospheric processes. *Environmental Science and Technology*, vol 22, p 361.

(6) EPA (1988) *Environmental Progress and Challenges; EPA's update*, EPA 230–07–88–033, US EPA, Washington, D.C.

(7) OECD (1991) *The State of the Environment—1991*, OECD, Paris.

(8) Uriarte, F.A. (1989) Hazardous Waste Management in ASEAN, in *Hazardous Waste Management* (eds S.P. Maltezon *et al*), Tycooly, London.

(9) Moore, J.N. and Luoma, S.N. (1990) Hazardous wastes from large-scale metal extraction. *Environmental Science and Technology*, vol 24, p 1278.

(10) Schweitzer, G.E. (1991) *Borrowed Earth, Borrowed Time. Healing America's Chemical Wounds*, Plenum Press, New York.

(11) Burns, P. (1988) Hazardous Waste Management—the way forward. *Journal of Institute of Water & Environmental Management*, vol 2, p 285.

(12) Deegan, J. (1987) Looking back at Love Canal. *Environmental Science and Technology*, vol 21, p 328.

(13) Yakowitz, H. (1989) Global hazardous transfers. *Environmental Science and Technology*, vol 23, p 510.

(14) UN (1989) *Illegal traffic in toxic and dangerous products and wastes,* Report of

the Secretary General A/44/362, United Nations, New York.

(15) Schneider, C. (1988) Hazardous Waste: the bottom line is prevention. *Issues in Science and Technology,* IV, p 75.

(16) Szenes, E. and Zoltai, N. (1988) The Hungarian experience in hazardous waste management. *Industry and Environment*, vol II, p 22.

(17) Postel, S. (1987) *Defusing the toxics threat,* Worldwatch paper 79, Washington, D.C., Worldwatch Institute.

Chapter 11

(1) FAO (1991) *The State of Food and Agriculture—1990*, FAO, Rome.

(2) WFC (1991) *Hunger and Malnutrition in the World*, World Food Council, Doc. WFC/1991/2, Rome.

(3) World Bank (1990) *World Development Report—1990*, Oxford University Press, Oxford.

(4) FAO (1991) *Current World Food Situation*, Document, CL/99/2, FAO, Rome.

(5) FAO (1991) *Livestock production and health for sustainable agriculture and rural development*, Background Document No 3, FAO/Netherlands Conference on Agriculture and Environment, FAO, Rome.

(6) FAO (1991) *Environment and sustainability in fisheries*, Document COFI/91/3, FAO, Rome.

(7) Bailey, C. and Skladany, M. (1991) Aquaculture development in tropical Asia. *Natural Resources Forum*, Feb. 1991, p 66.

(8) FAO (1991) *Sustainable development and management of land and water resources*, Background Document No 1, FAO/Netherlands Conference on Agriculture and Environment, FAO, Rome.

(9) Gale Johnson, D. (1984) World Food and Agriculture, in *The Resourceful Earth,* (eds J.L. Simon and H. Khan), Blackwell, Oxford.

(10) Revelle, R. (1984) The world supply of agricultural land, in *The Resourceful Earth*, (eds J.L.Simon and H.Khan), Blackwell, Oxford.

(11) Buringh, P. (1989) Availability of agricultural land for crop and livestock production, in *Food and Natural Resources,* (eds D. Pimentel and C.W. Hall), Academic Press, New York.

(12) FAO (1990) *FAO Production Yearbook*, vol 43, FAO, Rome.

(13) El-Hinnawi, E. (1991) *Sustainable agriculture and rural development in the Near East*, Regional Document No 4, FAO/Netherlands Conference on Agriculture and Environment, FAO, Rome.

(14) Repetto, R. (1986) *Paying the price: pesticide subsidies in developing countries*, Research Report No 2, World Resources Institute, Washington, D.C.

(15) Engelstad, O.P. (1984) Crop nutrition technology, in *Future Agricultural Technology and Resource Conservation*, (eds B.C. English *et al*), Iowa State University Press, Ames, USA.

(16) WHO (1990) *Public Health Impact of Pesticides Used in Agriculture*, WHO, Geneva.

(17) Pimentel, D. and Levitan, L. (1986) Pesticides amounts applied and amounts reaching pests. *BioScience*, vol 36, p 86.

(18) Conway, G.R. and Pretty, J.N. (1991) *Unwelcome Harvest*, Earthscan Publications, London.

(19) Crutzen, P.J. and Graedel, T.E. (1986) The role of atmospheric chemistry in environment-development interactions, in *Sustainable Development of the Biosphere,* (eds W.C. Clark and K.E. Munn), Cambridge University Press, Cambridge.

(20) Glotfelty, D.E. *et al* (1987) Pesticides in fog. *Nature*, vol 325, p 602.

(21) Rose, J.B. (1986) Microbial aspects of wastewater reuse for agriculture. *CRC Critical Reviews in Environmental Control*, vol 16, p 231.

(22) Enell, M. (1990) The impact on water quality of nitrogen losses from agriculture. *Acid Environment*, June 1990.

(23) WHO (1977) *Nitrates, Nitrites and N-Nitroso Compounds,* Environmental Health Criteria No 5, WHO, Geneva.

(24) Neal, R.A. (1984) Agriculture expansion and environmental considerations. *Mazingira*, July 1984, p 24.

(25) El-Hinnawi, E. and Hashmi, M. (1987) *The State of the Environment*, Butterworths, London.

(26) El-Hinnawi, E. (1991) Personal communication.

(27) El-Hinnawi, E. and Biswas, A. (1981) *Renewable Sources of Energy and Environment*, Tycooly International, Dublin.

(28) De Datta, S.K. (1986) Improving nitrogen fertilizer efficiency in lowland rice in tropical Asia. *Fertilizer Research*, 9, 171.

(29) Fullick, A. and Fullick, P. (1991) Biological pest control. *New Scientist*, Issues in Science, No 43, 9 March.

(30) *New Scientist*, 12 October 1991, p 16.

(31) FAO (1988) *Country Tables*, FAO, Rome.

(32) FAO (1990) *Fertilizer Yearbook*, vol 39, FAO, Rome.

(33) Trape, A.Z. (1985) The impact of agrochemicals on human health and the environment. *Industry and Environment*, vol 8, p 10, and in Georghiou, G.P. (1989) *Pest Resistance to Pesticides*, Plenum Press, New York.

Chapter 12

(1) WCED (1987) *Our Common Future*, World Commission on Environment and Development, Oxford University Press, Oxford.

(2) World Bank (1989) *World Development Report*, Oxford University Press, Oxford.

(3) World Bank (1990) *World Development Report*, Oxford University Press,Oxford.

(4) World Bank (1991) *World Development Report*, Oxford University Press, Oxford.

(5) UNIDO (1987) *Industry and Development, Global Report*, United Nations

Industrial Development Organization, Vienna.

(6) UNIDO (1990) *Industry and Development, Global Report*, United Nations Industrial Development Organization, Vienna.

(7) Bell, J. (1986) Caustic waste menaces Jamaica. *New Scientist*, 3 April, p 33.

(8) Kursten, M. *et al* (1988) Raw materials resources. *ATAS Bulletin*, vol 5, p 24.

(9) Ladou, J. (1984) The not-so-clean business of making chips. *Technology Review*, May/June, p 23.

(10) Postel, S. (1986) Increasing water efficiency, in *State of the World*, (eds L. Brown *et al*), W.W.Norton & Co., New York.

(11) OECD (1989) *Energy Balances of OECD Countries*, OECD, Paris.

(12) ECE (1991) *Energy Reforms in Central and Eastern Europe*, ECE Energy Series No 7, ECE, Geneva.

(13) World Bank (1986) *Industrial Energy Rationalization in Developing Countries*, John Hopkins University Press, Baltimore.

(14) Ross, M. (1989) Improving the efficiency of electricity use in manufacturing. *Science*, vol 244, p 311.

(15) Ross, M. and Steinmeyer, D. (1990) Energy for industry. *Scientific American*, vol 263, p 47.

(16) OECD (1990) *Energy Policies and Programmes of IEA Countries*, OECD, Paris.

(17) Flavin, C. (1986) *Electricity for a developing world*, Worldwatch Paper No 70, Worldwatch Institute, Washington, D.C.

(18) Jasiewicz, J. (1990) Comparison of the cost effectiveness of industrial energy conservation. *Natural Resources Forum*, Feb. 1990, p 70.

(19) El-Hinnawi, E. (1990) *Energy Conservation in Industry*, Report, National Research Centre, Cairo.

(20) OECD (1991) *The State of the Environment—1991*, OECD, Paris.

(21) Department of Environment (1989) *Digest of Environmental Protection and Water Statistics*, Her Majesty's Stationery Office, London.

(22) Hungarian Academy of Science (1990) *The State of the Hungarian Environment*, Budapest.

(23) *European Environmental Yearbook* (1987) DocTer International, London.

(24) Leonard, H.J. (1985) Confronting industrial pollution in rapidly industrializing countries. *Ecology Law Quarterly*, vol 12, p 779.

(25) Findly, R.W. (1988) Pollution control in Brazil. *Ecology Law Quarterly*, vol 15, p 1.

(26) Pimenta, J.C.P. (1987) Multinational corporations and industrial pollution control in Sao Paulo, Brazil, in *Multinational Corporations, Environment and the Third World*, (ed C.S. Pearson), Duke University Press, Durham.

(27) Castleman, B.I. (1985) The double standard in industrial hazards, in *The Export of Hazard: transnational corporations and environmental control issues*, (ed J.H. Ives), Routledge and Kegan Paul, Boston.

(28) Jassanoff, S. (1985) Remedies against hazardous exports: compensation,

products liability and criminal sanctions, in *The Export of Hazard: transnational corporations and environmental control issues,* (ed J.H. Ives), Routledge and Kegan Paul, Boston.

(29) Pearson, C.S. (1987) Environmental standards, industrial relocation, and pollution havens, in *Multinational Corporations, Environment and Third World* (ed C.S. Pearson), Duke University Press, Durham.

(30) Castleman, B.I. (1987) Workplace health standards and multinational corporations in developing countries, in *Multinational Corporations, Environment and Third World* (ed C.S. Pearson), Duke University Press, Durham.

(31) WHO (1990) *The Impact of Development Policies on Health*, World Health Organization, Geneva.

(32) WHO (1991) *Report of the Industry Panel*, WHO Commission on Health and Environment, WCHE/IND/2/7, WHO, Geneva.

(33) El-Batawi, M.A. and Husbumrer, C. (1987) Epidemiological approach to planning and development of occupational health services at a national level. *International Journal of Epidemiology*, vol 16, p 288.

(34) Nogueira, D.P. (1987) Prevention of accidents and injuries in Brazil. *Ergonomics*, vol 30, p 387.

(35) Frosch, R.A. and Gallopoulos, N.E. (1989) Strategies for manufacturing. *Scientific American*, vol 261, p 94.

(36) El-Hinnawi, E. (1991) Personal communication.

(37) OECD (1990) Energy and Technological Change. *STI Review*, No 7. OECD, Paris.

(38) El-Hinnawi, E. (1991) Personal communication.

Chapter 13

(1) World Bank (1991) *World Development Report*, Oxford University Press, Oxford.

(2) El-Hinnawi, E. (1981) The promise of renewable sources of energy, in *Renewable Sources of Energy and the Environment*, (eds E. El-Hinnawi and A. Biswas), Tycooly International, Dublin.

(3) World Resources Institute (1986) *World Resources—1986*, Basic Books, New York.

(4) Goldemberg, J. *et al* (1987) *Energy for a Sustainable World*, World Resources Institute, Washington, D.C.

(5) OECD/IEA (1991) *Greenhouse Gas Emissions: the energy dimension*, OECD/IEA, Paris.

(6) United Nations (1990) *Global Outlook—2000*, United Nations, New York.

(7) Davis, G.R. (1990) Energy for planet earth. *Scientific American*, vol 263, p 21.

(8) IAEA (1990) International Data File. *IAEA New Features*, No 8, September 1990, Vienna; *IAEA Bulletin*, vol 32.

(9) IAEA (1991) Nuclear power status around the world. *IAEA Bulletin*, vol 23, p 43.

(10) Semenov, B. *et al* (1989) Growth projections and development trends for nuclear power. *IAEA Bulletin*, vol 31, p 6.

(11) OECD/IEA (1989) *Energy and the Environment, Policy Overview*, OECD, Paris.

(12) UNEP (1981) *Environmental Impacts of Production and Use of Energy*. Tycooly International, Dublin.

(13) UNEP (1982) *The World Environment, 1972–1982*, Tycooly International, Dublin.

(14) UNEP (1984) *Comparative Assessment of the Environmental Impacts of Energy Production and Use*, Energy Reports Series, UNEP, Nairobi.

(15) El-Hinnawi, E. (1980) *Nuclear Energy and the Environment*, Pergamon Press, Oxford.

(16) El-Hinnawi, E. and Biswas, A. (1981) *Renewable Sources of Energy and Environment*, Tycooly International, Dublin.

(17) El-Hinnawi, E. *et al* (1983) *New and Renewable Sources of Energy*, Tycooly International, Dublin.

(18) El-Hinnawi, E. and Hashmi, M. (1987) *The State of the Environment*, Butterworth, London.

(19) OECD (1983) *Environmental Effects of Energy Systems*, OECD, Paris.

(20) OECD (1985) *Environmental Effects of Electricity Generation*, OECD, Paris.

(21) OECD (1989) *Emission Controls in Electricity Generation and Industry*, OECD, Paris.

(22) Chadwick, M.J. *et al* (1987) *Environmental Impacts of Coal Mining and Utilization*, Pergamon Press, Oxford.

(23) UNEP (1985) *Radiation: doses, effects and risks*, UNEP, Nairobi.

(24) UNSCEAR (1988) *Sources, Effects and Risks of Ionizing Radiation*, United Nations Scientific Committee on the Effects of Atomic Radiation, United Nations, New York.

(25) Gardner, M.J. *et al* (1990) Results of case control study of leukaemia and lymphoma among young people near Sellafield nuclear plant in West Cumbria. *British Medical Journal*, vol 300, p 423.

(26) Zhu, J.L. and Chan, C.Y. (1989) Radioactive waste management: world overview. *IAEA Bulletin*, vol 31, p 5.

(27) OECD (1991) *The State of the Environment—1991*, OECD, Paris.

(28) IAEA (1990) Decommissioning nuclear facilities. *IAEA News Features*, No 6, Feb. 1990.

(29) Holdren, J.P. (1990) Energy in transition. *Scientific American*, vol 263, p 109.

(30) El-Hinnawi, E. (1991) Personal Communication.

(31) OECD/IEA (1990) *Energy Policies and Programmes of IEA Countries*, OECD, Paris.

Chapter 14

(1) UNEP (1991) *Environmental Data Report*, 3rd edn, Blackwell, Oxford.

(2) Renner, M. (1988) *Rethinking the Role of the Automobile*, Worldwatch Paper 84, Worldwatch Institute, Washington, D.C.

(3) Faiz, A. *et al* (1990) *Automative Air Pollution*, The World Bank, Washington, D.C.

(4) United Nations (1979) *World Statistics in Brief*, United Nations, ST/ESA/STAT/SER V/4, United Nations, New York.

(5) United Nations (1990) *World Statistics in Brief*, 13th edn, United Nations, New York.

(6) OECD (1991) *The State of the Environment—1991*, OECD, Paris.

(7) OECD/IEA (1990) *Substitute Fuels for Road Transport*, OECD, Paris.

(8) Barde, J. and Button, K. (1990) *Transport Policy and the Environment*, Earthscan Publications, London.

(9) ECE (1991) *Energy Reforms in Central and Eastern Europe*, ECE Energy Series No 7, ECE, Geneva.

(10) UNEP (1987) *Kenya: National State of Environment,* Reports Series No 2, UNEP, Nairobi.

(11) El-Hinnawi, E. and Hashmi, M. (1987) *The State of the Environment,* Butterworth, London.

(12) Ndiokwere, C.L. (1984) A study of heavy metal pollution from motor vehicle emissions and its effects on roadside soil, vegetation and crops in Nigeria. *Environmental Pollution* (Series B), vol 7, p 35.

(13) Springer, K.J. (1982) Diesel emissions: a worldwide concern, in *Toxicological Effects of Emissions from Diesel Engines*, (ed J. Lewtas), Elsevier Science Publishers, Amsterdam.

(14) McClellan, R.O. (1987) Health effects of exposure to diesel exhaust particles. *Annual Reviews of Pharmacology and Toxicology*, vol 27, p 279.

(15) Egli, R.A. (1990) Nitrogen oxide emissions from air traffic. *Chimia*, vol 44, p 369.

(16) OECD/IEA (1991) *Greenhouse Gas Emissions: the energy dimensions*, OECD, Paris.

(17) Walsh, M.P. (1990) Motor vehicles and global warming, in *Global Warming,* (ed J. Leggett), Oxford University Press, Oxford.

(18) Bleviss, D.L. and Walzer, P. (1990) Energy for motor vehicles. *Scientific American*, vol 263, p 55.

(19) OECD (1985) *The State of the Environment—1985*, OECD, Paris.

(20) ECMT (1990) *Transport Policy and the Environment*, European Conference of Ministers of Transport, OECD, Paris.

(21) Foegen, J.H. (1986) Contaminated water. *The Futurist*, March/April, p 22.

(22) Chandler, W.V. (1985) Increasing energy efficiency, in *State of the World*, (ed L. Brown), W.W. Norton, New York.

(23) El-Hinnawi, E. (1991) Personal Communication.

(24) Lowe, M.D. (1990) *Alternatives to the Automobile: transport for livable cities*, Worldwatch Paper 98, Worldwatch Institute, Washington, D.C.

Chapter 15

(1) WEFA (1989) *The Contribution of World Travel and Tourism Industry to the Global Economy*, a study for American Express Travel, New York, Wharton Economic Forecasting Associates.

(2) Grenon, M. and Batisse, M. (1989) *Futures for the Mediterranean Basin*, Oxford University Press, Oxford.

(3) Ascher, F. (1985) *Tourism: transnational corporations and cultural identities*, UNESCO, Paris.

(4) Dixon, J.A. and Sherman, P.B. (1990) *Economies of Protected Areas*, Earthscan Publications, London.

(5) Harcourt, A.H. *et al* (1986) Public attitudes to wildlife and conservation in the Third World. *Oryx*, vol 20, p 152.

(6) Morris, H. and Romeril, M. (1986) Farm tourism in England's Peak National Park. *The Environmentalist*, vol 6, p 105.

(7) French Ministry of Environment (1987) *State of the Environment—1987*, Paris.

(8) Jackson, I. (1986) Carrying capacity for tourism in small tropical Caribbean islands. *Industry and Environment*, vol 9, p 7.

(9) El-Hinnawi, E. (1986) *Energy Conservation in Buildings*, Symposium on Energy Conservation, Energy Planning Agency, Cairo.

(10) Salm, R.V. (1986) Coral reefs and tourist carrying capacity: the Indian Ocean experience. *Industry and Environment*, vol 9, p 11.

(11) Michaud, J.L. (1983) *Le Tourisme Face à l'Environnement*, Presses Universitaires de France, Paris.

(12) WTO (1990) *Economic Review of World Tourism*, World Tourism Organization, Madrid.

Chapter 16

(1) World Resources Institute (1986) *World Resources—1986*, Basic Books, New York.

(2) United Nations (1989) *World Population Prospects—1988*, ST/ESA/SER.A/106, United Nations, New York.

(3) UNICEF (1990) *Children and Development in the 1990s*, UNICEF, New York.

(4) UNICEF (1988) *The State of the World's Children—1988*, Oxford University Press, Oxford.

(5) WCED (1987) *Our Common Future*, World Commission on Environment and Development, Oxford University Press, Oxford.

(6) UNEP (1982) *The World Environment, 1972–1982*, Tycooly International, Dublin.

(7) Simon, J.L. (1981) *The Ultimate Resource*, Princeton University Press, Princeton.

(8) Simon, J.L. and Khan, H. (1984) *The Resourceful Earth*, Blackwell, Oxford.

(9) MacNeill, J. (1989) Strategies for sustainable economic development. *Scientific American*, vol 261, p 105.

(10) UNFPA (1991) *Population and the Environment*, United Nations Population Fund, New York.

(11) Morris, D. (1979) *Measuring the Condition of the World's Poor: the physical quality of life index*, Pergamon Press, New York.

(12) Camp, S.L. and Speidel, J. (1987) *The International Human Suffering Index*, Population Crisis Committee, Washington, D.C.

(13) UNDP (1990) *Human Development Report—1990*, United Nations Development Programme, Oxford University Press, Oxford.

(14) UNDP (1991) *Human Development Report—1991*, United Nations Development Programme. Oxford University Press, Oxford.

(15) McNamara, R.S. (1981) *The McNamara Years at the World Bank*, John Hopkins University Press, Baltimore.

(16) World Bank (1990) *World Development Report—1990*, Oxford University Press, Oxford.

(17) UNICEF (1989) *The State of the World's Children—1989*, Oxford University Press, Oxford.

Chapter 17

(1) UNEP (1982) *The World Environment, 1972–1982*, Tycooly International, Dublin.

(2) United Nations (1989) *World Population Prospects—1988*, United Nations, New York.

(3) UNFPA (1991) *Population and the Environment: the challenge ahead*, United Nations Population Fund, United Nations, New York.

(4) El-Hinnawi, E. (1991) *Sustainable Agriculture and Rural Development in the Near East*, Regional Document No 4, FAO/Netherlands Conference on Agriculture and Environment, FAO, Rome.

(5) Harpham, T. and Stephens, C. (1991) Urbanization and health in developing countries. *World Health Statistics Quarterly*, vol 44, p 62.

(6) WHO (1989) *Spotlight on the Cities: improving urban health in developing countries*, World Health Organization, Geneva.

(7) UNDP (1991) *Human Development Report—1991*, United Nations Development Programme, Oxford University Press, Oxford.

(8) Hardoy, J.E. and Satterthwaite, D. (1989) *Squatter Citizen: life in the urban Third World*, Earthscan Publications, London.

(9) Cointreau, S.J. *et al* (1984) *Recycling from Municipal Refuse*, World Bank Technical Paper No 30, World Bank, Washington, D.C.

(10) Lohani, B.N. (1984) Recycling potentials of solid waste in Asia through organized scavenging. *Conservation and Recycling*, vol 7, p 181.

(11) Furdey, C. (1984) Socio-political aspects of the recovery and recycling of urban wastes in Asia. *Conservation and Recycling*, vol 7, p 167.

(12) Hardoy, J.E. *et al* (1990) *The Poor Die Young*, Earthscan Publications, London.

(13) Kingman, S. (1991) South America declares war on chagas disease. *New Scientist*, 19 October.

(14) WHO (1984) *Biomass Fuel Combustion and Health*, EFO/84.64, World Health Organization, Geneva.

(15) Smith, K.R. (1986) Biomass combustion and indoor air pollution. *Environmental Management*, vol 10, p 61.

(16) Chen, B.H. *et al.* (1990) Indoor air pollution in developing countries. *World Health Statistics Quarterly*, vol 43, p 127.

(17) Saliba, L.J. and Helmer, R. (1990) Health risks associated with pollution of coastal bathing water. *World Health Statistics Quarterly*, vol 43, p 177.

(18) Mayo, S.K. *et al* (1986) Shelter strategies for the urban poor in developing countries. *Research Observer*, vol 1, p 183.

Chapter 18

(1) Needelman, H.L. *et al* (1990) The long-term effects of exposure to low doses of lead in childhood. *The New England Journal of Medicine*, vol 322, p 83.

(2) Demaeyer, E. and Adiels-Tegman, A. (1985) The prevalence of anaemia in the world. *World Health Statistics Quarterly*, vol 38, p 302.

(3) UNICEF (1989) *Strategies for Children and Development in the 1990s*, Executive Board Paper, UNICEF, New York.

(4) WHO (1987) *Evaluation of the Strategy for Health for All By the Year 2000*, World Health Organization, Geneva.

(5) Hofvander, Y. *et al* (1981) Organochlorine contaminants in individual samples of Swedish human milk. *Acta Paediatrica Scandinavia*, vol 70, p 3.

(6) Slorach, S.A. and Vax, R. (1983) *Assessment of human exposure to selected organochlorine compounds through biological monitoring*, Swedish National Food Administration, Uppsala.

(7) Jensen, A.A. (1983) Chemical contaminants in human milk. *Residue Review*, vol 89, p 1.

(8) Karakaya, A.E. *et al* (1987) Organochlorine pesticide contaminants in human milk from different regions of Turkey. *Bulletin of Environmental Contamination and Toxicology*, vol 39, p 506.

(9) Kocturk, T. and Zetterstrom, R. (1988) Breast feeding and its promotion. *Acta Paediatrica Scandinavia*, vol 77, p 183.

(10) Lopez, A.D. (1990) Causes of death: an assessment of global patterns of mortality around 1985. *World Health Statistics Quarterly*, vol 43, p 91.

(11) Haaga, J. *et al* (1985) An estimate of the prevalence of child malnutrition in developing countries. *World Health Statistics Quarterly*, vol 38, p 331.

(12) WHO (1989) Mortality and morbidity: global estimates. *World Health Statistics, Annual Report—1989*, WHO, Geneva.

(13) WHO (1985) World malaria situation—1983. *World Health Statistics Quarterly*, vol 38, p 193.

(14) WHO (1990) World malaria situation—1988. *World Health Statistics Quarterly*, vol 43, p 68.

(15) Hunter, J.M. *et al* (1991) *Parasitic Diseases in Water Resources Development*, WHO Document (in preparation).

(16) Talla, I. *et al* (1990) Outbreak of intestinal schistosomiasis in the Senegal River Basin. *Annales Société Belge Médecine Tropical*, vol 70, p 173.

(17) UNICEF (1990) *The State of The World's Children*, Oxford University Press, Oxford.

(18) Kurzel, R.B. and Cetrulo, C.L. (1981) The effect of environmental pollutants on human reproduction, including birth defects. *Environmental Science and Technology*, vol 15, p 626.

(19) Kalter, H. and Wakary, J. (1983) Congenital malformations. *The New England Journal of Medicine*, vol 308, p 423.

(20) Shane, B. (1989) Human reproductive hazards. *Environmental Science and Technology*, vol 23, p 1187.

(21) WHO (1990) *Public Health Impact of Pesticides Used in Agriculture*, WHO, Geneva.

(22) Stevens, J.B. and Swackhamer, D.L. (1989) Environmental pollution: a multimedia approach to modeling human exposure. *Environmental Science and Technology*, vol 23, p 1180.

(23) Lioy, P.J. (1990) Assessing human exposure to airborne pollutants. *Environmental Science and Technology*, vol 25, p 1361.

(24) IARC (1989) *Biennial Report*, International Agncy for Research on Cancer, WHO, Geneva.

(25) Esrey, S.A. *et al* (1985) Interventions for the control of diarrhoeal diseases among young children: improving water supplies and excreta disposal facilities. *Bulletin World Health Organization*, vol 63, p 757.

(26) UNICEF (1989) *The State of the World's Children*. Oxford University Press, Oxford.

(27) Hirschhorn, N. and W.B. Greennough (1991) Progress in oral rehydration therapy. *Scientific American*, vol 264, p 16.

Chapter 19

(1) SIPRI (1986) *World Armaments and Disarmament*, SIPRI-Yearbook 1986, Oxford University Press, Oxford.

(2) SIPRI (1988) *World Armaments and Disarmament*, SIPRI-Yearbook 1988, Oxford University Press,Oxford.

(3) SIPRI (1990) *World Armaments and Disarmament*, SIPRI-Yearbook 1990, Oxford University Press,Oxford.

(4) World Bank (1988) *World Development Report—1988*, Oxford University Press, Oxford.

(5) UNDP (1990) *Human Development Report—1990*, United Nations Development Programme, Oxford University Press, Oxford.

(6) UNDP (1991) *Human Development Report—1991*, United Nations Development Programme, Oxford University Press, Oxford.

(7) United Nations (1989) *Study on the Economic and Social Consequences of the Arms Race and Military Expenditures*, Disarmament Study Series 19, United Nations, New York.

(8) SIPRI (1984) *Herbicides in War*, Taylor and Francis, London.

(9) UNEP (1991) *Report on the UN Inter-Agency Plan of Action for the ROPME Region. Phase I*, UNEP OCA/PAC, UNEP, Nairobi.

(10) Price, A.R.G. and Sheppard, C.R. (1991) The Gulf: past, present and possible future status. *Marine Pollution Bulletin*, vol 22, p 222.

(11) Hahn, J. (1991) Environmental effects of the Kuwait oil field fires. *Environmental Science and Technology*, vol 25, p 1531.

(12) SIPRI/UNEP (1985) *Explosive Remnants of War*, Taylor and Francis, London.

(13) El-Hinnawi, E. (1985) *Environmental Refugees*, UNEP, Nairobi.

(14) Ehrlich, P.R. *et al* (1983) Long-term biological consequences of nuclear war. *Science*, vol 222, p 1293.

(15) Turco, R.P. *et al* (1983) Nuclear winter. *Science*, vol 222, p 1283.

(16) Turco, R.P. *et al* (1984) The climatic effects of nuclear war. *Scientific American*, vol 251, p 33.

(17) Ehrlich, P.R. (1984) Nuclear winter. *Bulletin Atomic Scientists*, April 1984, p 3S.

(18) Grover, H.D. (1984) The climatic and biological consequences of nuclear war. *Environment*, vol 26, p 7.

(19) Covey, C. *et al* (1984) Global atmospheric effects of massive smoke injections from a nuclear war. *Nature*, vol 308, p 21.

(20) United Nations (1985) *Climatic Effects of Nuclear War, Including Nuclear Winter*, Report A/40/440, United Nations, New York.

(21) National Research Council (1985) *The Effects on the Atmosphere of a Major Nuclear Exchange,* National Academy Press, Washington, D.C.

(22) Svirezhev, Y.M. (1985) *Ecological and Demographic Consequences of Nuclear War*, USSR Academy of Science, Computer Centre, Moscow.

(23) SCOPE (1985) *Environmental Consequences of Nuclear War. Vol II,* SCOPE Report No 28, J. Wiley, Chichester.

(24) SCOPE (1986) *Environmental Consequences of Nuclear War. Vol I,* SCOPE Report No 28, J. Wiley, Chichester.

(25) Dotto, L. (1986) *Planet Earth in Jeopardy*, J. Wiley, Chichester.

(26) Peterson, T. (1986) Scientific studies of the unthinkable—the physical and biological effects of nuclear war. *Ambio*, vol 15, p 60.

(27) Goldblat, J. (1975) The prohibition of the environmental warfare. *Ambio*, vol 4, p 187.

(28) Barnaby, F. (1976) Towards environmental warfare. *New Scientist*, 69, 6.

(29) SIPRI (1977) *Weapons of Mass Destruction and Environment*, Taylor and Francis, London.

(30) SIPRI (1984) *Environmental Warfare*, Taylor and Francis, London.

(31) United Nations (1982) *The Relationship Between Disarmament and Development*, E82.IX, United Nations, New York.

(32) United Nations (1983) *Economic and Social Consequences of the Arms Race and of Military Expenditures*, E83.IX.2, United Nations, New York.

(33) United Nations (1985) *Study on Conventional Disarmament*, E85.IX.1, United Nations, New York.

(34) United Nations (1989) *Disarmament Series No 19*, United Nations, New York.

(35) United Nations (1986) *Concepts of Security*, Disarmament Series No 14, United Nations, New York.

(36) SIPRI/UNEP (1986) *Global Resources and International Conflict,* Oxford University Press, Oxford.

(37) Renner, M. (1989) *National Security: the economic and environmental dimensions*, Worldwatch Paper No 89, Worldwatch Institute, Washington, D.C.

(38) Renner, M. (1990) *Swords into Plowshares: converting to a peace economy*, Worldwatch Paper No 96, Worldwatch Institute, Washington, D.C.

(39) SIPRI/UNEP (1985) *Explosive Remnants of War*, Taylor and Francis, London.

(40) El-Hinnawi, E. (1991) Personal Communication.

Chapter 20

(1) Solocombe, D.S. (1984) Environmentalism: a modern synthesis. *The Environmentalist*, vol 4, p 281.

(2) Hart, S.L. (1980) The environmental movement: fulfillment of a renaissance prophecy? *Natural Resources Journal*, vol 20, p 501.

(3) Passmore, J. (1978) *Man's Responsibility for Nature*, Duckworth, London.

(4) O'Sullivan, P.E. (1986) Environmental science and environmental philosophy. *International Journal of Environmental Studies*, vol 28, p 97.

(5) Capra, F. (1983) *The Turning Point: science, society and the rising culture*, Flamengo, London.

(6) Morrison, D. (1980) The soft cutting edge of environmentalism. *Natural Resources Journal*, vol 20, p 275.

(7) O'Riordan, T. (1981) *Environmentalism,* Pion, London.

(8) UNEP (1988) *The Public and the Environment*, UNEP, Nairobi.

(9) CEC (1986) *The Europeans and their Environment in 1986*, Commission of the European Communities, Brussels.

(10) Johnson, B.B. (1987) Public concerns and the public role in siting nuclear and chemical waste facilities. *Environmental Management*, vol 11, p 571.

(11) Environment Agency of Japan (1982) Public opinion poll on environmental pollution. *Japan Environment Summary*, vol 10, p 1.

(12) OECD (1987) *Environmental Data Compendium*, OECD, Paris.

(13) OECD (1991) *The State of the Environment—1991*, OECD, Paris.

(14) Harris/UNEP (1988) *Public and Leadership Attitudes to the Environment in Four Continents*. Louis Harris & Associates, New York and UNEP, Nairobi.

(15) EPA (1987) *Unfinished Business: a Comparative Assessment of Environmental Problems*, US Environmental Protection Agency, Washington, D.C.

(16) Wilkins, L. (1987) *Shared Vulnerability: media coverage and public memory of the Bhopal disaster*, Greenwood Press, Westport, Conn.

(17) Wilkins, L.and Patterson, P. (1987) Risk analysis and the construction of news. *Journalof Communication*, vol 37, p 80.

(18) Friedman, S.M. *et al* (1987) Reporting on radiation: a content analysis of Chernobyl coverage. *Journal of Communication*, vol 37, p 58.

(19) Sood, R. *et al.* (1986) How the news media operate in natural disasters. *Journal of Communication*, vol 37, p 27.

(20) Adams, W.C. (1986) Whose lives count ? TV coverage of natural disasters. *Journal of Communication*, vol 36, p 113.

(21) Gaddy, G.D. and Tanjong, E. (1986) Earthquake coverage by western press. *Journal of Communication*, vol 36, p 105.

(22) Downs, A. (1972) Up and down with ecology—the issue attention cycle. *Public Interest*, vol 28, p 38.

(23) WCED (1987) *Our Common Future*, World Commission on Environment and Development, Oxford University Press, Oxford.

(24) Elkington, J. and Hailes, J. (1988) *The Green Consumer Guide*, Victor Gollancz, London.

(25) George, M.K. (1989) Seeing the green light. *South*, September, 1989.

(26) Kellner, J. (1990) Beware of the greencon. *New Internationalist*, No 203, January 1990.

(27) Shea, C.P. (1989) Doing well by doing good. *World Watch*, November/December, Washington, D.C.

(28) Collison, R. (1989) The greening of the boardroom. *Business Magazine*, July 1989.

Chapter 21

(1) Nicholson, M. (1987) *The New Environmental Age*, Cambridge University Press, Cambridge.

(2) EPA (1987) *EPA's Use of Cost-Benefit Analysis*, US Environmental Protection Agency, Washington, D.C.

(3) OECD (1991) *The State of the Environment—1991*, OECD, Paris.

Chapter 22

(1) UNEP (1982) *The World Environment 1972-1982*, Tycooly International, Dublin.

(2) UN (1971) *Development and Environment.* Report submitted by a panel of experts convened by the Secretary-General of the United Nations Conference on the Human Environment, Founex, Switzerland, 4-12 June

1971, Kungl. Boktryckeriet, PA Norstedt and Soner, Stockholm.

(3) UN (1973) Report of the United Nations Conference on the Human Environment, United Nations, New York. A/CONF.48/14/Rev.1

(4) UN (1974) The Cocoyoc Declaration adopted by the participants in the UNEP/UNCTAD Symposium on 'Patterns of Resource Use, Environment and Development Strategies' held at Cocoyoc, Mexico from 8 to 12 October 1974, United Nations, New York. A/C.2/292

(5) UNEP (1980) *Choosing the Options; Alternative Lifestyles and Development Options*, UNEP, Nairobi.

(6) UN (1980) *Development and International Economic Co-operation*, United Nations, New York. A/35/592/Add.1

(7) UNEP (1974) Introductory Report by the Executive Director to the second session of the Governing Council, UNEP, Nairobi. UNEP/GC/14

(8) Tolba, M.K. (1974) Statement to the World Food Conference, Rome, November 1974, in M.K. Tolba: *Development without Destruction*, Tycooly International Dublin, 1982.

(9) UNEP (1975) Report of the Governing Council on the work of its third session, UNEP, Nairobi.

(10) Brown, B.J. et al (1987) Global sustainability: Toward definition, *Environmental Management*, vol 11, p 713.

(11) WCED (1987) *Our Common Future*, Oxford University Press, Oxford

(12) UNEP (1987) *Environmental Perspective to the Year 2000 and Beyond*, UNEP, Nairobi.

(13) Lee, J.A. (1985) *The Environment, Public Health and Human Ecology: Considerations for Economic Development*, John Hopkins University Press, Baltimore.

(14) Pearce, D. et al (1989) *Blueprint for a Green Economy*, Earthscan Publications, London.

(15) OECD (1991) *The State of the Environment 1991*, OECD, Paris.

(16) Saunders, R.J. and J.J. Warford (1976) *Village Water Supply*, World Bank, Washington DC.

(17) Barbier, E.B. (1987) The concept of sustainable economic development, *Environmental Conservation*, vol 14, p 101.

(18) Repetto, R. et al (1989) *Wasting Assets: Natural Resources in the National Income Accounts*, World Resources Institute, Washington DC.

(19) World Bank (1991) *World Development Report 1991: The Challenge of Development*, OUP, New York.

(20) SIPRI (1991) *The SIPRI Yearbook 1991 World Armaments and Disarmament*, Oxford University Press.

(21) UNDP (1991) *Human Development Report 1991*, Oxford University Press.

(22) UNEP (1984) *The State of the Environment 1984*, UNEP, Nairobi.

(23) The Langkawi Declaration on Environment (1989) *Development and International Economic Co-operation: Environment*, UN, New York. A/44/673

(24) The South Commission (1990) *The Challenge to the South*, Oxford University Press, Oxford.

(25) IUCN/UNEP/WWF (1980) *World Conservation Strategy*, IUCN, Gland.

(26) IUCN/UNEP/WWF (1991) *Caring for the Earth: a Strategy for Sustainable Living*, IUCN, Gland.